Writing on Computers in English Comp

Second Edition

Goran "George" Moberg

Borough of Manhattan Community College
City University of New York

Published in association with

The Writing Consultant
P.O. Box 20244
New York, NY 10025

KENDALL/HUNT PUBLISHING COMPANY
2460 Kerper Boulevard P.O. Box 539 Dubuque, Iowa 52004-0539

Other books by Goran "George" Moberg:
Writing in Groups
How to Teach Writing in Groups [co-author]

This edition has been printed directly from camera-ready copy.

Library of Congress Catalog Card Number: 89–64103

ISBN 0–8403–5753–2

Printed in the United States of America
10 9 8 7 6 5 4 3 2 1

"O that my tongue were in the thunder's mouth!
Then with a passion would I shake the world."
--Shakespeare

Don't get it right.
Get it written.
Then get it right.

"He put a new song in my mouth."
--Psalms

Write first.
Edit later.

TRADE TOOLS ALWAYS IN THE WRITER'S BAG:

<><><><><><><><><><><><><><><><><><><><><><><><><><><><><><><>

- PENS
- YELLOW PADS
- NOTEBOOOK FOR JOURNAL
- **WRITING ON COMPUTERS**
- DICTIONARY
- FLOPPY DISK
- ____________________

CONTENTS

PREFACE to the New Edition

One of the most popular features of this text during the three-year tenure of the first edition has been the illustration of persuasive writing through real examples of the way good students write; in fact, because of many requests for even more sample writings by students, this new edition includes about twice as many. There's also a section about talking and writing about fiction. Other improvements: since many schools now have the Realtime Writer in their computer system, this edition includes some suggestions for using this fascinating program; the order of the chapters has been rearranged to reflect the typical order of assignments, as reported to me by kind users of the book; finally, some minor imperfections have been removed.

The first day I taught a computer-writing class I overheard a couple of students talking about word processing: they had seen secretaries press a few keys on an office machine, only minutes later to pull out finished letters and reports. "I'm so glad I won't have to write in this English class," one of these students said. "Yes," said the other one, "word processing will do the writing for us."

Of course, this book is not about word processing. It's about writing. It's a composing manual. That's why its title is not "Word Processing in the English Classroom." Automated secretarial work has little to do with composing one's own writing. In fact, expert operators of word-processing equipment often do not know how to write. Conversely, authors do not say that they "word-process" (except when they work temporary office jobs); the phrase I do hear them use, though, is "writing on my computer."

Writing on Computers is a basic rhetoric (with plenty of sample student essays), designed for high school and college English courses. If you are a teacher, all you need to use this book in a computer-writing class is computers (with printers) and a program for writing. That's it. You don't need any of the outlining, heuristics, vocabulary, or grammar-drill programs now coming on the market. You might choose to add one or more of them, but they are not necessary. I like to keep things simple, so my students use the computer simply to write on, as a supplemental, labor-saving, and exciting variant of the old pen-and-paper way of writing. And the students don't need prior experience with computers--or even with typewriters, though some typing experience is helpful.

Drafts of this book have been class-tested by half a dozen teachers. Our aim has been to develop a process in which computer-writing enhances the longhand of the students, their longhand in turn improving their work on the machine in a continuing felicitous circle.

Three chapters (14, 15, 16) train the students in voicing their "self-editing." This successful "editing-aloud" method for tracking down and fixing grammatical and mechanical departures from standard English has proved particularly effective when students edit on computer screens and print-outs.

Throughout Writing on Computers I recommend the use of groups and "collaborative learning." My experience--confirmed by the writings of colleagues in the professional literature (e.g. James Collins, Stephen Marcus, Dawn Rodriguez, Helen Schwartz, Cynthia Selfe)--shows that computers and groups go together naturally and effectively. The groups humanize the machine; the machine refines the technology of composing, editing, and publishing. Thus my text first explains the formation of groups and later presumes their presence.

Although Writing on Computers grew out of my own classroom practice, I've been much inspired and have learned plentifully from mentors, colleagues, and students. Some of them I know personally, others I know only from their writings or lectures. Naturally, none of them is responsible for any shortcomings in my work. I hasten to say, too, that not all of them use computers--some of them shun these machines. But this is not a computer book. It's a writing book. I'm very grateful for all the help and encouragement from these gracious teachers and writers:

Milton Baxter, James Berlin, Bill Bernhardt, Ann Berthoff, Lynne Beyer, Pat Bizzell, Robert Boynton, Kenneth Bruffee, Joel Climenhaga, Phil Eggers, Peter Elbow, Susan Fawcett, David Fletcher, Brian Gallagher, Lee Gurshuny, Carolyn Kirkpatrick, Ron Lieber, Harry Lutrin, Roberta Matthews, Peggy Dye Moberg, James Moffett, Don Mulligan, Skai Picciotto, Yvonne Stam, John Trimbur, George Warner, and Harvey Wiener.

Thanks also to John Short (The Writing Center), to James Tynes (Learning Resource Center), Tom Lew (Data Processing), and to Sandra Poster (Office of Academic Affairs)--all of the Borough of Manhattan Community College. I'm also indebted to the Dover Pictorial Archives for the Gustave Dore and Thomas Nast illustrations. And special thanks to Tony Drago.

I have also learned greatly from my students. I am truly grateful to them all, but especially to those who allowed me to publish their writings in this book: Alex Augustus, Ursula Bailey, Maxine Berman, Carrie Bolden, Eugenia Cartagenia, Margaret Cox, Andrea Davis, Magaly Familia, Fernando Garcia, Mary Ingram, Veronica Jackson, Gilbert Johnson, Marie Johnson, Irina Khlevner, Donald Macaluso, Dolores Payan, Tanya Pringle, William Purcell Jr., Miguel Reyes, Chitreka Singh, Stan Styk, Evelyn Velez, John Williams, and Tien Xuan Ho.

December, 1989

G. M.

Intermission--Sample Student Voice:

Dolores Payan / English 090-513 / December 9, 1985

The Computer and My English Course

When I heard for the first time that I would be writing on a computer in my English class, I thought, "Oh, my God, something terrible will happen to me." I didn't like writing. And I had never touched a computer before.

The first day when I sat in front of the computer, I felt strange and scared. The teacher and Tracy, a young girl who worked in the lab, explained to us how to write on it. My impression was that we were wasting our time. I felt we should learn how to write English instead of fooling with these machines.

I have a big problem with the computer. It comes when I have to insert a word at the end of a line. I always press the wrong key and erase half a sentence, or the whole paragraph rocks back and forth. But I've now learned to control this.

We help each other in the class. Earlier I used to call on the teacher or on Tracy when I needed help with the machine. But now Paulette and Elizabeth, who are in my group and usually sit on either side of me, give me help every time I need them. They have more experience with computers than I have.

The computer has helped me to correct some of my writing problems. I draft the essay on my yellow pad. I rewrite and edit it at home. Then again I revise it and edit it on the screen in the lab before I print it. The computer has also helped me to find grammar errors. I see clearly when the verbs are in the wrong tense, when the "s" is missing on plural nouns, or when I have a run-on sentence. My handwriting is such a mess that sometimes I can't read what I have written. But on the computer screen the writing is clear and I can seee most of my mistakes.

During the process of practicing my writing on the computer in my English class I have learned to like writing. What a change. Earlier I used to hate it. Now I'm pleased when the teacher writes on my paper, "Excellent--file it." He means the essay is right and I don't have to do it over.

Last week he set aside one computer for feedback. He put one student's essay on the screen. We all then had the chance to read it and type in comments on the computer. This was so interesting that some of us went back later and gave feedback on the others' feedback.

I'm glad the school has introduced the computer program in some of the English courses. I believe the computer will help other students who have writing problems the same way it has helped me.

"When are we going to get the computers, Ma'am?"

Read These Questions and Answers First:

Ω Ω Ω Ω Ω

Question: Do I have to learn about electronics to write on computers?

Answer: No. But you'll want to learn to use the keyboard right away. The sooner you feel comfortable with the keys, the sooner you'll write up a storm.

Question: Yes, the keyboard--that's what I thought. I'll have to be a perfect typist.

Answer: No, that's not necessary at all. If you're already a pretty good typist, you'll probably learn to write on the computer a little faster than if not. But hunting and pecking with two fingers will do just fine for good writing on a computer. In fact, if you've never typed, you might not make the mistake that typists often make when they first come to computers: they unnecessarily hit the carriage return at the end of every line.

Question: I see. But what happens if I hit the machine too hard? Won't it break?

Answer: Don't worry! Your equipment is sturdy, built to take a lot of beating. If your computer should malfunction--all machines are liable to break down occasionally--it's not your fault. (If your computer should stop working suddenly, however, could it be something simple and easy to fix?)

"This computer isn't helping my writing at all."

1 / INTRODUCTION TO COMPUTER-WRITING AND TO GROUP-WRITING

Doing a lot of work at a machine can be a lonely job, as people in factories have always known. Along with recent advances in technology it has also been discovered that people are most productive and happy when working collaboratively in small groups.

And indeed, it has been my experience as an English teacher that students learn to write on computers most effectively when their class is divided into workshop-like groups. Not that the writing itself has to be collaborative, but dealing with the **process of writing**--chiefly giving and getting **feedback**--is done most wonderfully in small groups, with or without computers.

Let's talk about the computers first and then the groups. By the way, I usually advise my students to use a personal voice in their writing--and I try to practice what I teach--so in this book you'll hear my voice speaking directly to you.

Now first I want you to visualize the following scene. Imagine that you're looking for a job as a management trainee at the office of Hi-Tek Employment Co. You've just handled a tough interview and come through with flying colors, and now you're being led into a new room, and--lo and behold--it's full of computers.

"The next thing we want you to do," says the personnel agent, "is to write a short essay about your career goals. Use that computer by the window. There's paper already in the printer."

And you go to it with gusto because you've taken a computer-writing course and know exactly what to do. Lucky you--and you get the job, too!

I presented you with that scene **not** to sing a hymn to the role of electronics in our society, nor to drive home what a great aid computers can be in helping people improve their writing skills (though that is certainly true). No, I have a totally different point in mind. Since your ability to write on a computer was **assumed** in that example, the scene was of course set in the future. How far in the future? Ten-twenty years? Who knows how long it will be before most educated people can do their writing on computers. To get to my point, let's visualize another scene, also set in the near future:

Mr. Wizzy Futura, 20 years old, has just finished a series of tests and interviews at Old-fashioned Employment Inc., and only one task remains--writing a short essay about his career goals. The person who is screening Wizzy hands him a clipboard with some white paper and a ball pen. "Here, give me about 250 words."

"Bbbut where's your word processor?" says our hero. You see, ever since grade school he has done all his writing on the latest computer equipment.

"Oh, we've got one," says the agent of Old-fashioned Employment Inc., "but it broke down and we haven't fixed it yet. Just use pen and paper. Here."

The fact, however, is that Wizzy hasn't used a pen for years and really wouldn't know how to write an essay without a computer. Naturally, he didn't get the job. As he left the office with downcast eyes, he recalled, too, how he had quarrelled painfully with his former girl friend when she was away at college. He had sent her love letters that were dot-matrix print-outs and looked like office memos. "I want some love scrawls from you, dripping with ink, like blood and tears," she had written him on perfumed paper. "Write to me like a human being, not like a machine," she had begged him. She was married to somebody else now.

You get the drift. This is a course in **writing,** not in word processing, not in computer expertise. You'll be writing both on your yellow pad with an old-fashioned pen **and** on a computer. Back and forth. One method will reinforce the other: your work on the computer will improve your longhand, your longhand will enhance your computer-writing, all in a felicitous circle.

Let me add one strong, immediate reason why you ought not give up your longhand: even if you'll be allowed to take the final in this writing course on a computer, in how many other courses can you use computers for your essay tests and term papers? We don't give up walking because of technological advances in the means of transportation. Actually, as you come to discover your own best writing process, you may find that you prefer to do some of your drafting and revision with an old pencil stub, sometimes on a yellow pad, sometimes on a computer print-out.

Now let's see how the computer will benefit your writing. You may already be convinced that it will speed your progress. But if you're skeptical or scared of this new technology, let me reassure you: most of the research findings now coming in--as well as the subjective impressions--report good news. And I feel the same way.

- Beginners learn computer-writing fast.
- The screen and the print-outs are so clear and easy to read that most students fix errors more easily than before.
- Studies indicate that most students develop their essays more effectively than before they wrote on computers.
- Surveys show that a majority of students **enjoy** computer-writing. This increases their motivation.

ooOOOOOoo

Let's now turn from the electronic to the human environment--your classmates and your teacher. That's where the groups come in.

GROUP-WRITING:

In most cases, a computer-writing class will meet some of the time in a classroom--without computers--and the rest of the time in some sort of **computer lab.** So the idea is that you regroup in the lab, staying always in the same group. But before I further explain the method and the benefits of group work, I want to tell you a little story:

A couple of years ago some benevolent citizens of an affluent suburb decided to do something good for the pre-school children from less affluent homes in their community. So they invited the children to come every day throughout the summer and learn to swim in various private pools of the suburb. At the end of the summer, a study found that about one third of the children had learned to swim.

Amazingly, however, it turned out that **all** the children who went to one particular pool had mastered the art of swimming. When the organizers of the pool program investigated, they found that this pool had been the only one with heated water. Evidently the kids enjoyed the warm water and therefore learned to swim faster. A warm learning environment!

Similarly, it's known that most students are ill at ease learning to write within the chilly walls of a typical classroom. It doesn't feel comfortable for writing. And no matter how pleasant and skilled a teacher you have, nothing beats your own getting together with your peers in a close-knit group to create just the kind of warm and friendly environment that you need to develop your writing skills in comfort. That's the genius of the group way: you help create your own learning environment. Yet at the same time your teacher is still in charge, offering guidance when you need it.

The gist of the system is rich flexibility: each group becomes both a private little workshop and a part of a larger, semi-public unit, the class. At the same time, your teacher--freed from some supervisory chores--is able to hold one-to-one conferences right in the classroom, as in a lab. But when desired, of course, your teacher may convene the whole class for lecture or discussion. So you get the best of three worlds in this setup:

- You're a member of your **group;**
- you're a member of a twosome **conference** with your teacher whenever your need personal coaching; and
- you're a member of your whole **class.**

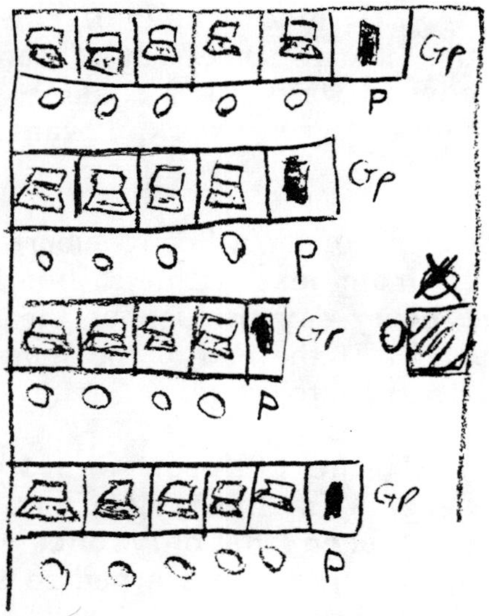
Gp
P
Gp
P
Gr
P
Gp
P

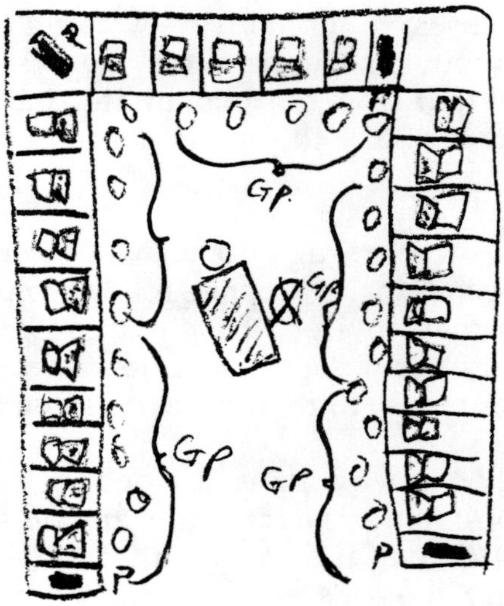
Gp
Gp
Gp
P
P

P
P
P
P

WRITING IN GROUPS

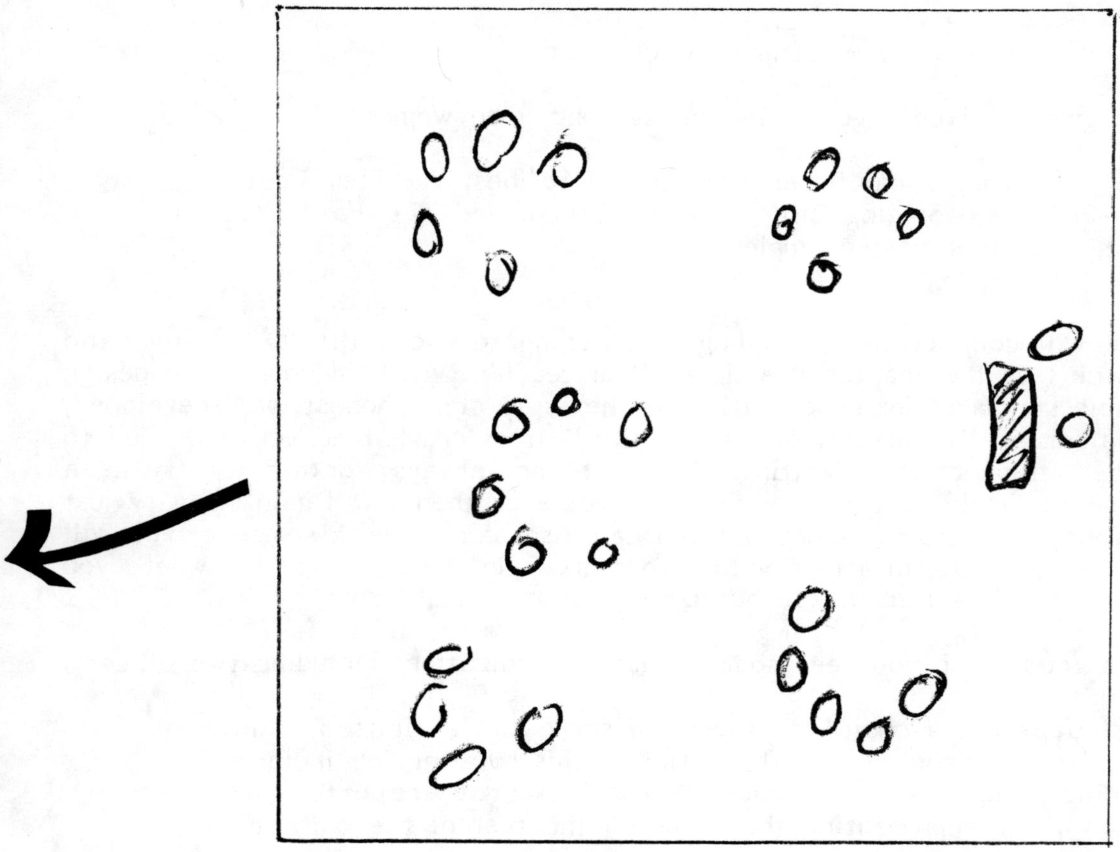

Above you see a good floorplan for groups. It has worked smoothly for years in classrooms across America. Please note the horse-shoe configuration: in this way--with a slight swivel or twist of the chair or the neck--all student-writers can face the teacher up front (by the X). Or they can face the center of their own group, as the choice may be. A typical class will change back and forth between the "class mode" and "group mode," though not often or abruptly.

The computer lab, on the other hand, works best in the "group mode" all the time, the teacher seeing individual students in conference either by their machines or at a desk. The horse-shoe shape usually cannot be retained in the lab, but the groups should stay together and keep their identity as well as possible. On the left half of this spread you can see some typical layouts of computer labs. The most common one is the upper left one with straight rows: each row would be one group. Writing on a computer could be a lonely job, so stay in close touch with your group mates.

So let's get started. If you haven't grouped yet, move your chairs to form horseshoe-shaped groups [see the charts on the last two pages!] The ideal size is five. But four or six members are fine, too. Ask your teacher if there are any rules or suggestions. I recommend the following:

- The first few sessions you may switch groups.
- Keep groups mixed (age, ethnic background, men/women).
- Give your group a catchy name (Tribe of Scribes, The Four Poets, Pen-Fighters, Singing Birds, Honest Group, and The Fantastic Five are some past examples).

Often you will collaborate by reading or listening to each other's writings and offering **feedback** (see the chapter this title). Your teacher will help your group design a set of desirable standards for good writing: is the essay clear, honest, well developed, enjoyable, grammatically correct, and so forth. Writers crave feedback and need to know what others think of their writing. But you're not always working jointly as a group. At times you simply write, all by yourself, yet even then you'll gain security just by sitting in your little group. Sometimes you can also pair off with someone; you will be especially likely working in a pair within your group in the computer lab where you probably will not sit in a semi-circle but perhaps in a straight row.

To have a good group you need some minimal organization, including two officers:

- A **chairperson** ("chair" for short) to serve for a limited time, say, two or three weeks. The duties of this rotating job includes leading groups discussion, appointing a daily **group reporter** (see below), and representing the group to the rest of the class and the teacher.
- A **group reporter**, who serves one day at a time, takes notes during the session, both in the classroom and in the lab, and then writes a **Group Report** for your teacher (see the chapter with that name!).

If you've been in groups before, you know how well they work. But if not, you may be concerned about showing your writing to strangers who are going to criticize your personal opinions. I certainly sympathize with that apprehension, but listen to this: First, you won't be strangers for long. Second, tell the others in your group that you don't want any criticism. You want praise! Tell them that you'd rather be ruined by praise than saved by critique. Actually you'll be surprised at the friendly and constructive comments you'll receive.

For that matter, you don't have to show your writing to anybody--until you're willing and ready. All you have to do for now is to sit in a group and do some writing, first on a yellow pad and then on a computer.

MY BILL OF RIGHTS

1. I, ______________________, have the right to write on my computer at my own speed and to feel OK if I'm slower than someone else.

2. I have the right to ask for help with my computer. And to get it.

3. I have the right to be imperfect at my keyboard.

4. I have the right to do a sloppy first draft.

5. I have the right to feel good about myself even if I make a mess on my screen.

6. I have the right, too, of not being in love with my computer.

7. I have the right to be treated like an adult by lab attendants, teachers, and peers.

8. I have the right to be proud of my own print-outs regardless of what others might think of them.

9. I have the right to write all I want.

NOTES ABOUT MY COMPUTER

2 / USING THE MACHINERY

YOUR EQUIPMENT:
The Computer, the Writing Program, the Printer

Since this is a course in **writing,** and not in data processing, you'll get only a minimum of technical information about the machinery in your lab or classroom. In the same way, it's important for you to realize that this is **not** a secretarial course in word processing, but rather a course in **how to write good essays.** The computer simply serves as an effective tool for writers.

So what do we mean with the term "computer"? Though in some schools students work at a "terminal" connected to a "network" or a big "mainframe," in most places today students are using a "micro" (small) or a "personal" computer, also known as a "PC". But to keep things simple, this book will assume that your machine is a micro or a PC. [Use your wide MARGIN to the right to note the specific, minor differences that come up if your school uses a network or a mainframe.]

[In some of the chapters in this book you'll find an extra-wide right margin, like here. This space is for your comments on the text to the left and for additional instructions.

[Sometimes your teacher may suggest what to write here. Or you yourself may decide to insert some notes that will help you later on. For example, if the book tells you about the <u>**ERASE**</u> function, you may want to jot down which keys to press to <u>**ERASE**</u> text.

As you saw in the heading at the top of this page, your machine is really three things:

1. your **computer** (micro or PC) = hardware

2. your writing **program** (on a disk) = software

3. your **printer** = more hardware

1. The **computer,** too, has three components:

a) **the "keyboard"**--which you type on more or less like a typewriter.

b) **the "screen"** or **"monitor"**--which looks more or less like a TV.

c) **the "drives"**--where you insert your **"disks".**

Now please turn the page so we can get down to the real point of all this: how do we **write** on the computer?

What really happens when you "write" on a computer?

You create your writing by hitting the **letter** keys. The "brain" of the computer reflects your keyboard action by putting your words on the screen where they stay for you to fool around with until you turn the machine off. Then everything you wrote is erased from the computer's memory.

If you want to save your text, however, you can do that by transferring it from the memory behind the screen onto a disk in the drive below. Then you pull the **disk** out and take it with you, just like a yellow pad that you write on. Next time you use your computer, you put the disk back in the drive, and your writing shows up on the screen again. In a nutshell, that's what happens. In a little while, you're going to try it out yourself.

Before we go on, two words of caution:

a) Do not touch any switches on the equipment until you've received proper training from your teacher or lab attendant.

b) Handle your **disk** with kid gloves. This means that you keep it either in the disk drive or in its sleeve in a safe place. There are many things you should **not** do with it--see the illustration to the right [courtesy & copyright 1983, Beagle Bros Inc., publisher of Apple II software]. Remember, if you spill coffee on the disk that holds your term paper, you don't have a term paper any more. That's why eating or drinking is not allowed in most computer rooms.

What you should do when handling your disk is always to hold it by its label--label facing up--your thumb on the label.

2. The writing program ("word processing"). This is the program that makes your computer a writing tool rather than, say, an accounting or engineering tool. Some of these programs are simple, some complex, but they are all fairly easy to learn yet powerful enough to give you all you need to write essays, revise them, and print them. I used the word "powerful" because I believe that--as soon as you get the hang of your program--you'll feel you've gained a new power as a writer.

In the beginning of the course, you'll probably find that, when you arrive, a lab attendant or your teacher has already put the writing program into your computer's memory, so you won't have to **boot** the program, as it's called. But later on, you'll probably learn how to do your own booting.

ALWAYS SAVE YOUR WRITING

at the End of Each Work Session!

3. **THE PRINTER:** If your computer is not hooked up to a printer, you'll have to do the following: To print your writing, you'll first save it, of course, on your data disk.

To **SAVE** your daily writing, you'll need your own **data disk** (as contrasted with the "program disk"). After performing the saving command, you take your **disk** out of its drive and bring it to one of the printers in the lab. In the beginning the lab attendant or your teacher might do the printing for you. But later they'll show you how to do your own printing.

Even before you see improvement in your writing, you'll be delighted with how good it looks when printed out.

IT'S EASIER THAN YOU THINK

Intermission--Sample Student Voice:

Magaly Familia / English 090 / Dec. 11, 1985

How the Computer Helped Me Fix My Grammar Mistakes

I was absent the first day, so it was hard for me when I came to the second class. And it had been a long time since I had used a keyboard. Beside that, my nails were so long that I couldn't type a word. But the following session I came prepared--I had cut my nails to a reasonable length.

At first the lab was just a place where I would put on the computer screen whatever piece of writing I'd done in class before the lab. At that early time, I didn't do much to improve my writing. I would give my paper to the teacher who would find all kinds of things I could do to improve it. He put some remarks on it and I'd go back to the computer, make changes, and then give him a clean copy. Yet I didn't put anything of my own in the changes.

As anyone could see, I wasn't doing much. I was just writing sort of carelessly. After a while that changed, however. I started being more responsible for my work. I knew I had to pass this course. Besides, the teacher changed his system of marking. He was no longer finding the mistakes for us. He was just putting a dot in the margin next to the line that had a mistake. Whenever he returned a paper to me, I would take some time to find my mistakes and correct them. In some cases I couldn't find the mistake in the line with the dot. That's when I turned to my group mates. They always helped me when I was in need. If we couldn't solve the problem together, that's when I asked the teacher.

After I had everything fixed on the paper, I would enter the changes on the screen. I would read the text out aloud to check it, and then once again on the new print-out. At this point I would return my essay to the teacher. And wouldn't you know it--he'd still find mistakes, but not as many as before.

This computer lab has helped me a lot because when I go there I have the opportunity to look at my writing more closely than when I only have my handwriting. This is a second chance to see what I didn't see when I wrote the first draft with pen. It helps me feel responsible and creative to see my own work in a different light, the green light of the screen. Each day I look forward to a print-out with fewer and fewer errors.

When I started this course my main problem was with grammar. I used to make a great deal of mistakes with verb forms and noun plurals. I used to put the verb in the present when it should be in the past, and vice versa. I would put the plural "s" on a singular noun. Using the screen and the print-outs as well as the reading-aloud method of our text, I'm now getting sharp at catching these things.

3 / YOUR FIRST WRITING SESSION AT THE COMPUTER

1. **Start writing.** Go slow and easy. Even if you have had some experience with keyboards and feel these beginning actions are too simple for you, please follow the instructions exactly, step by step. Soon enough you'll be able to advance quickly.

a) Write the word **george** by pressing the keys for each letter.

b) Since that first g doesn't look right, let's change it by making it a CAP (capital). First, to keep it simple, erase or delete the whole word by pressing the "delete" key [**DEL**] six times. Have you found it? It's the key that erases your letters backwards.

c) Next you locate the two SHIFTKEYS--each lower side. Press one of those keys and hold it down while with your other hand you press G. Now let go of the SHIFTKEY and continue to type out eorge.

d) But you don't want George on **your** screen--you want to write **your** name. Therefore you now use the **DEL** key and erase George and then write in your own name. If you make a mistake in typing, just use the **DEL** key to erase and then do it over. Next you press the long SPACEBAR at the bottom of the keyboard to make a space before you type your last name. Do you know how to get a CAPITAL letter now? If you need help, ask your neighbor, lab attendant, or teacher.

e) Next you want to move down one line, so you press the **CR** key [CARRIAGE RETURN]. Even though your computer has no carriage (as on a typewriter), most people call it that. It's an important key, sometimes called just the "return" or the "enter" key as well. You'll find the **CR** to the right.

I just hit the CR key now. Can you see that it interrupted the previous line and gave us a new line? So you can use it **to end a paragraph.**

I just used it again--TWICE. What happened? In addition to a new line, the computer gave me a **blank line** as well! But on your machine, please press the CR just **once,** to get ready for the next line.

2. Next, **using also the number keys in the top row,** type out English 100-123 (course name). Have you found the **hyphen** key? Finally press the CR.

3. Now that you're getting warm, typing today's DATE on line 3 should be child's play. First type Sept. 15, 1986 (for practice), and then delete it and instead put in today's correct date. Then hit CR twice.

What you have on your screen now should look something like this:

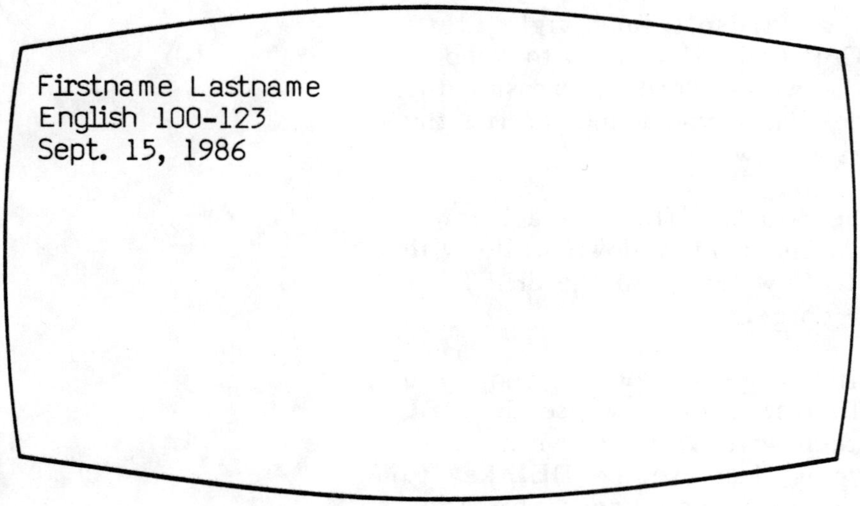

OK? TAKE YOUR TIME. There's no rush. Some people are faster than others, and that's all right. If your group mate is faster than you, ask him or her for help. Or ask your teacher or lab attendant.

In a few moments, you'll copy a short passage. In the process you'll learn about WORDWRAP and about how to use the CURSOR. First the cursor.

When you're typing just one line (the way you have been doing it so far), it's easy to correct an error by deleting it backwards with the DEL key. But what if you've already typed many perfectly good lines--except for one mistake on the first line? Do you have to erase everything? Not at all. Use the cursor:

4. The **CURSOR.** As you've already seen, that blinking little bug on your screen (I call it glowworm) marks the spot where your letters show up as you type them. Don't curse your cursor but learn to love it! The reason is that your cursor is a writing tool, totally under your control. It will help you shape your writing. Let me explain.

Basically we do only two things when we write, whether on a computer or on paper: a) we put words down, and then b) we **change** them. The simplest way to change your writing on a screen is to press the DEL key to rub out the last letters or words that you wrote. You already know how to do that.

But let's say you want to go back several lines or a whole paragraph to change a word--how's that done? The answer is that you move the CURSOR to the spot where you want to change something. Here's an example. Look at your own screen right now. The last thing your wrote was today's date. Then you hit the CR key twice, and you're about ready to begin copying that paragraph about wordwrap that I mentioned. But before you go on, let's imagine that you decide you want to add--we call it "insert" in this case--**your middle name** between your first and last names.

INSERT ON/OFF:

----- -----

[At this point, make sure your computer is in the **INSERT** mode. This means that when you type new letters on top of old ones, the old words are simply pushed away to the right. This is standard on most machines, but on some you can switch the INSERT mode ON and OFF.]

----- -----

CURSOR KEYS:

The next thing you do is move the CURSOR on top of the first letter of your last name. **USE THE CURSOR KEYS.** (Usually you'll find them on a key pad on the right.) If you look closely, you'll see tiny arrows on four different keys that show cursor movements **UP, DOWN, LEFT, RIGHT--one space or line at a time.**

Play with your cursor for a bit. Notice how it moves around freely like a ghost or an angel; it never erases or moves any text. When you get a feel for the cursor's motion, **leave it on the first letter of your last name.**

Now, go to the next page to do the insertion:

a) **INSERTION:** Now simply type your **middle name** plus a space. See how easy it was to make an insertion!

Just for the sake of practice, change your mind and delete (remove) your middle name.

Now put it back in. Writers keep changing their mind, you know. When you've decided how you want to leave your name, **move the cursor back down to where you were before.** You can do this with the four cursor keys we just talked about, but here's shortcut: your keyboard also has a "page down" key. Try it. How did it work out? **<u>Pg. Dn.</u> Key:**

b) **INDENT:** Before you start typing that passage about WORDWRAP that I promised some time ago, I want you to learn to INDENT for a new paragraph. Just press the **TAB** key (it's usually placed in the upper left corner). Hit it! What happened? **How to <u>INDENT</u>:**

5. Now here's the WORDWRAP piece. INDENT and start copying:

> This is the first time I have ever written on a computer. I can see that a word I type at the end of a line suddenly moves to the beginning of the next line automatically without my having to do anything about it. That's called wordwrap. It means that the words arrange themselves in lines, "wrapping" themselves from line to line. So I don't have to press the CR key (as on a typewriter) unless I want to end a paragraph--like here: NOW I press the CR.

c) **CORRECTIONS:** Now read aloud to see and hear if you copied correctly. If you find any typing errors, use your CURSOR and DEL keys to go back and correct them. So far I've told you about only one correction key, but some systems have **two** different keys for deleting. Find out if yours does, and if so, explain the difference in the margin to the right!

6. Next, copy the short fable, but first I want you to learn how to **CENTER.** so please ask your teacher or lab attendant and then write down how to do it in the margin to the right.

How to <u>CENTER</u>:

Now type and CENTER the title of the fable. And do you remember how to INDENT the first line? [TAB] Let wordwrap decide where your lines will end. Now copy:

> The crow felt lucky when she found a pitcher of water because she was very thirsty. The pitcher was only half-full of cool water, however, so when the crow tried to drink, she found her beak was too short. She tried again and again but in vain. Just as despair was about to make her give up, she got an idea. She began dropping pebbles, one by one, into the clear water. With each little stone, the water rose in the pitcher until it was high enough for the sly bird to reach the liquid and slake her thirst.
>
> MORAL: <u>Necessity is the mother of invention.</u>

How to <u>UNDERLINE</u>:

Did you manage to underline the moral? If not, find out how to do it, and then jot it down in the margin.

b) **What did you think of the fable?** Perhaps you were too busy learning how to use the machine to think about the fable itself. Well, let's discuss its moral another time. But right now, go back and correct some of the typing mistakes you might have made. Use the cursor and the DEL keys. I said **some** of the typing errors because I don't want you to become a perfectionist and get stuck. Make a few corrections and move on.

It's time now to look at some of the **non-writing functions:** how to start a new piece of writing, how to save it, how to get an old piece out on the screen, and a few other house-keeping chores.

A Few Computer-Writing Terms Explained:

FILE: Most of us are working on several pieces of writing simultaneously--a letter here, a report there, and so on. Those we're not dealing with at the moment are "filed" away somewhere, perhaps in a drawer or a briefcase. When you use a computer, you similarly store your separate writings in **files** on your data disk. Each **file** has a name, of course, so you'll know what is what. (On some systems, a file is called a "document.")

How to **NAME** a file:

File Name: If you haven't named your fable file yet (your first file), I suggest you call it BIRD, just to keep it simple. It doesn't matter what you call a file, except that it has to be short--usually 8 characters (sometimes you can add a period and three more characters). It may help your teacher if you standardize the file names of essay assignments in the course.

How to **SAVE:**

Let's say you're ready to quit for the day. To file away your file is called to **SAVE** it. Find out how SAVING is done on your equipment. Do it now! Save the BIRD file. Do it even if you're not ready to quit or to print. Do it for practice. Then write down the procedure in the right margin for your reference next time around in case you've forgotten.

Ho to **OPEN** OR **RETRIEVE:**

Now you want to get the BIRD file out again. [By the way, what's on this file? Three things: 1) the heading with your name, course, and date; 2) the passage about wordwrap; and 3) the crow fable.] There are several terms for retrieving a file from the disk so it appears on the screen. In fact, one of them is **RETRIEVE.** Another common term is **OPEN** a file. Put the term and the procedure in the space to the right.

Hold it--a disaster just happened.... Just as you got your BIRD file back on the screen, somebody walked behind your machine and accidentally rubbed against the electric cord so it came loose. Your screen went blank, of course, and your BIRD file is gone! Or is it?

Relax--no disaster at all. Remember this, every time you bring your writing from the disk onto the screen, **it's only a COPY.** The real thing stays in storage on the disk.

How to **CLEAR** the screen:

CLEAR the screen: Some writing programs automatically clear the screen when you close a file by SAVING your text to disk. But others don't. They just give you the screen message, "file saved," while your screen remains cluttered by a file you don't want. How does your system CLEAR the screen? (Answer in the margin.) By the way, if you unintentionally CLEAR your screen, you now understand that nothing is lost, except a few seconds of time; you simply retrieve the file you were working on from your disk.

How to **COPY**:

COPY a file. Occasionally we writers desire two copies of a piece of writing we're working on, perhaps to use one of the files as a **BACKUP** for extra safety (some systems do this automatically--does yours?). Another reason for using the COPY function to make a duplicate file could be to rewrite each one of the two in a different manner and then print them both out and compare them. If that sounds like a lot of work, try doing it with pen and paper! (Jot down in the margin the procedure for the COPY function.)

How to **ERASE** or **DELETE** a passage or file:

DELETE or **ERASE** a passage: This function we're dealing with here is **not** the familiar rubbing out of the last few letters or words, but the surgical removal of several lines or a paragraph or even a page or more. In fact, a related function is to **DELETE A FILE**--you destroy a whole file, as when you throw out a draft you don't need any more. What keys do you press to perform these kinds of surgery? (Explain in the margin.)

How to **MOVE** text:

MOVE text around: This sophisticated action for writers who want to get down to serious **revision.** It's a fabulous time-saver. Imagine crossing out a 15-line paragraph in one place and then writing it over again with your pen elsewhere? Without the computer, many of us would just as soon skip all that rewriting. But it's a piece of cake with the MOVE command. How do you do it with your system? (Put in margin.)

After all this reading, let's get back to computer work. On the next page we're going to try out some of the functions we just discussed. So now we want you to RETRIEVE or OPEN your BIRD file and get ready for action.

More Practice on the BIRD File:

How to divide a paragraph in two:

1) If you wanted to divide the paragraph about the thirsty crow into **two** paragraphs, exactly **where** would you do it? And **how** would you do it? Go ahead--do it now: Put the cursor on the first letter of the word that you want to start the second paragraph. Hit the CR. Then hit the TAB. That's it! (It works this way only if you're in the INSERT mode, of course.) Explain in the margin the two steps you take to divide one paragraph into two.

How to join two paragraphs:

2) As we noted earlier, writers are fickle and often change their mind. Those two paragraphs now look awfully short, don't you think? Put them back together so we have only that one paragraph! This is how you do it: put the cursor on that same first letter of the new paragraph and gently press the DEL key (the one that moves **backwards** as it deletes). Press it gently several times until the paragraphs are joined again. Wasn't that nifty?

How to eliminate a line:

3) Design change in your heading: bring your CURSOR back up to **your name** [press **"home"** or **"page up"** cursor key]. I recommend that, from now on, you **begin** every file you write in the same way: Instead of keeping your name, course, and date on three separate lines, you put them all on **one** line, like this:

Your Name / English 100-001 / March 1, 1986.

Did you figure out how to do it? [You pull each line up with the DEL key, the way you closed up the paragraphs.] By the way, ask your teacher exactly how you ought to design your "title" or "heading." For example, "Essay #x" may be included.

4. The next change involves practicing the **ERASE** function. I want you to take out that first paragraph about wordwrap. We don't need it anymore, now that you understand wordwrap. Can you do this on your own? Try. You'll be using the ERASE function often in the future.

5. Now you're ready to SAVE and PRINT your BIRD file. Then you can check the "hard copy" of this practice file. Finally, instead of copying, you're ready to begin composing your own writing

So, **congratulations** for having come this far already. You see, you're now going to move away from just **copying** somebody else's text (that's word processing) and to start **creating your own writing.** You will either compose at the keyboard or enter words on the screen from your handwritten draft. Who knows what music will come out of this!

Before we end this long technical chapter, I just want to ask you a question about the process of your writing. When you transfer your own text from your yellow pad onto the screen, do you have to **copy** it exactly, the way you copied the wordwrap and thirsty-bird passages?

I would suggest "no" as an answer. In the earlier assignment, you practiced mechanical skills, to learn the operation of the keyboard. But from now on you're going to do **creative** work. So if you think of a better way to say what you want to say than the way you did it on your yellow pad, then it would make sense to go ahead and change it. (Does your pad have to be yellow? No. It's just a tradition for writers. And I think it's a good idea to use yellow paper for drafts and white for finished copy.)

Do you remember a while back that you were asked what you thought of the fable you copied, the **meaning** of the story and its moral? Well, we'll deal with that and similar issues of meaning in the chapter called "General and Specific."

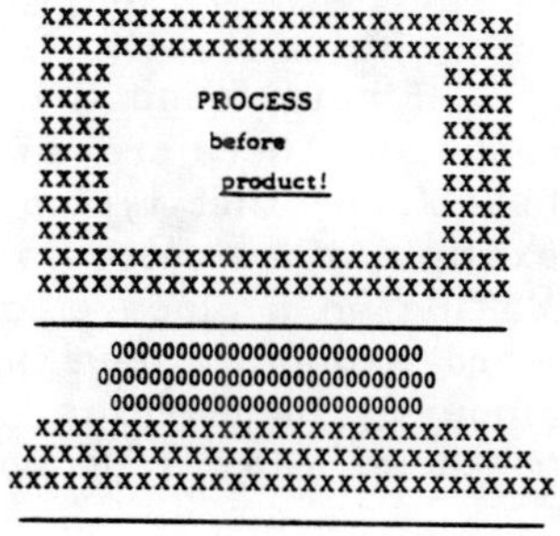

P.S. Just one more technical point: Always **SAVE** your writing on your disk before you take it out of the drive. Always.

Intermission--Sample Student Voice:

John Williams / English 090-513 / December 11, 1985

THE COMPUTER AND MY WRITING

On the day of registration I heard about the prospect of doing remedial writing on the computer. I had to make an on-the-spot decision. After weighing the alternatives and reviewing my past experiences in English classes, I came to the conclusion that writing on the computer would be better in the long run. That is, I would improve my writing skills and at the same time be able to use a computer. I can say with confidence that I can now write better than before and also that I am able to operate a word-processing program.

I always enjoyed going to this writing class. The computer has been a stimulant to keep me from getting bored in the class and with the writing I do. Last semester my writing sometimes turned to boredom, especially when there was nothing interesting for me to write. But not so in this class! Everytime I looked up from my computer screen, I could see my classmates deep in concentration. They were trying to complete one of their essay assignments.

It seems that whenever I start to work on the computer, time passes swiftly. I would be in the middle of an essay, as the teacher would say, "You only have fifteen minutes to print." In times like these I get excited. Before ten minutes of the allotted time were up I had to finish printing what I had typed so that I could edit the print-out. It is from the print-out that I am able to see my grammatical mistakes and get a good idea of what I'm saying about the chosen topic.

We all helped each other in my group. When I wanted to insert or delete a few sentences I would call on my neighbors to read what I had written. They would tell me if they understood what I had written and whether there was room for improvement. If I mistrusted their opinion, I would call over the teacher and rely on his judgement. This process helped me to develop my own style of writing.

Because I can add or delete sentences at will on the computer, I can make a paragraph short or long. This is very helpful when I begin writing the first paragraph of an essay. It was here my greatest weakness lay. A good opening paragraph will help me to write a good essay. Since using the computer I have been able to solve this problem.

When I have spare time, I do experiments which include searching for new commands and finding faster ways to delete or add paragraphs or sentences. There are so many things I would like to do at once when I sit before the computer, such as, reviewing and completing past essays, finding new commands, completing present assignments, and improving my typing speed. I cannot do all these at once so I do the most important which is the current assignments.

I have found that since I have been using the computer, my editing errors have been greatly reduced. There are fewer run-on sentences, transpositions of letters, and incorrect word endings. I have found that typing the word letter-by-letter helped me concentrate on what I was doing. For example, if I wanted to type the word "but" I would call each letter to myself before I typed them. Writing on a piece of paper I will do the same. Because of this I can readily tell when I write a word wrong. I have also found that my tendency to write disjointed sentences has been reduced. Since my handwriting has improved, however, the computer print-out has given me a sense of neatness. Now I do not like to see finished essays with messy, scratchy corrections.

There were many times early in the course when I punched the wrong command key and did not know what to do. Whenever situations like these occurred, I would become excited and confused. Sometimes I would call the teacher for help, or one of the lab attendants. Other times I would try to solve the problem on my own. If I was able to solve the problem, I felt I had accomplished something. Writing on the computer is fun to me and a powerful learning experience.

4 / JOURNAL* MEANS DAILY

March 5 '86

...For the last couple of days I've been writing promotional material, not just writing but designing some flyers for an author's radio-show scouting program operating out of California. My agent asked me to join it for publicity, so now I'm making up these flyers on the hot issue of writing in American education. Perhaps I'll be on a few radio shows.

Composing for publicity is different from text-booook writing, requires another tone and vocabulary. Although it's hard to imagine my audience--could be anybody--I'm fascinated with this promotional work, especially the page-layout. Actually, that's a big part of what paragraphing is all about, making a pleasing page to look at for my readers. Some people forget this about paragraphs and talk only of "unity" and "coherence" as if people had ***no eyes.***

It's a bit scary that this work is so much fun that it may be dangerous for my book writing. (Should I get a job on Madison Avenue?) I have to watch how I spend my time because it's easy to fritter away a whole day fooling with different formulations and different page-layouts. Like right now I'm itching to play with different typefaces for the HEADLINE....

I need to write the next chapter in the book instead, Your Learning Log or something like that. Maybe I'm not ready. I hesitate. I can't decide on the chapter title. I'm intrigued by our choice of word when we fire up our computers for a session of composing and say that we "log on" to a drive. I have a nice feeling for that curt word--log--making me think of adventures on the high seas. ["Today when the storm finally subsided, a red sun appeared through the mist along with a mysterious ship...."] I also wrote a student brochure for a college once, "Maintaining a Learning Log."

Yet everybody now is into JOURNAL-WRITING. That's the thing, that's what they call it. I don't want to use log just to be different. Well, good-bye, I'd better get back to my book again.

---

*Etymology: **journal** is a French loan word, original sense "daily."

That was an entry from my journal on the previous page. I write about my writing (and other topics) every morning on my computer and save it to disk. I used to do this in notebooks. The oldest one says "1948" on the cover. You could say it's a habit for me to write in my "daily book." I still make journal-like entries on scraps of paper and in the margins of magazines. If I don't lose them, I later put these jottings on disk. I believe such saved writings are valuable capital for one's writing. So in my classes I do not recommend journal-writing: I make it a requirement.

Whether you call your notebook **journal, log,** or **diary** is not important. The dictionary does not support those who claim that diary means one thing and journal another.... Here's one definition of **journal:** "A daily record of occurrences, as a diary or ship's log." So don't bother to argue with people who insist that a diary is personal and a journal objective, or a journal emotional and a diary not. Your own daily writing book becomes what you make of it.

How to <u>REPLACE:</u>

I'm going to use the term journal here, but I'll confess something: when I first wrote this chapter, I used the word "log" throughout. Then I changed my mind, as is our prerogative as writers. But thanks to the word-processing function REPLACE (also called Search & Replace or Find & Replace), I made the change in a jiffy: I just hit a few keys and in a flash every log had become a journal. I hope you'll soon learn the **replace** function on your machine because it makes revision easier.

"How's the journal different from automatic writing?" is a common question:

> The JOURNAL is a **genre** and a physical object, a **book** in which you write every day, whereas AUTOMATIC WRITING is a technique (fast, non-stop writing). But many people choose to practice automatic writing in their journal.

THE SPIRIT OF THE JOURNAL:

Genre means "type of literature," so you'd put journals alongside poems, plays, stories, and essays as one special kind of writing. While some authors have produced great journals, our object in this course is not literary art but simply the practice of a form.

For most people, the more **personal** your writing the more you develop your own writing skills. Your journal is a private place where you can relax and be yourself, perhaps find yourself. It's a private **record** where you put down anything from sudden bright ideas to the high price of coffee, if that's important to you; from boring hours through angry incidents to moments of bliss; from intellectual problems to economic and romantic affairs; and from last night's real dream through this afternoon's day-dream to your writing problems. Aha--finally I came to the point:

> Though ultimately you'll write whatever you want to, I have found that those students have benefited the most who have made their journals into **writer's journals**, meaning that whenever possible they made their own thoughts, feelings, and ideas about their progress as writers into their primary subject.

So I suggest that every day when you open your journal, you **first** see if you want to make any comment on how your writing is coming along. You'll be amazed many times --I can almost guarantee you--that once you begin discussing with yourself the difficulties you're encountering in your writing tasks, for example, your own response in your journal-writing will bring solace, solutions, and sometimes even a great opening for an essay that's due in just a few days--right there in the journal. So guard it with your life.

To keep your juices flowing **freely** as you write, remember that your journal will **not** be criticized or corrected. Not all teachers handle this matter the same way, but here's what I do: I never read my students' journals unless they specifically ask me and permit me to look at certain entries. But what I do, of course, is check them periodically (without actually reading them) to see that there are **daily** entries.

Once a student was so proud of a journal entry that she asked me to publish it in our class bulletin. Other students from time to time pass their journals around in their groups--while some keep them under lock and key. It's up to you.

THE PHYSICAL JOURNAL:

As an object, the journal is usually a notebook, whether a 49¢ plain-wrapper one or a fancy hardcover book. I've seen them come leather-bound, even. It's important that you select one that you'll be comfortable with, and better a large one than one so little that you have to buy another later when you run out of space. Note that it should be a **separate** book--not a pad, nor a section of your class notes, nor some sheets loose in your bag. Treat your journal-notebook with respect and love.

"Group reports, automatic writing--and now a diary, too. This sounds like an awful lot of writing for just one course. Is all this necessary?" some of you may be thinking at this point.

You will not make progress if all the writing you do is the finals and the "big" essays. Yes, "all this" is necessary. The reason many people have problems with their writing is that it's something they do as a task separate from everything else **once in a while,** only when absolutely required. Look at it this way: writing is like loving--the expression of your entire body and soul. Could you imagine suddenly turning on your ability to love about three times a month **on command,** if you never practice it in-between? Your journal keeps you warm between your important essay assignments and also prepares you for them.

Some students are delighted when they learn of the journal assignment because they already keep one and love it. Others think of it as hardship. If you belong to the second category, let me tell you that many reluctant writers have been transformed by the new experience of starting a journal. Perhaps this will happen to you. Your journal might turn into your best friend and confidant.

ON THE COMPUTER:

If you have easy access to a computer, you could try and see how you'd like making journal entries at the keyboard. Make sure you put the print-outs in your journal-book, for I wouldn't recommend that you do your journal-writing exclusively on the screen. Perhaps you remember the concern I expressed in the first chapter that society might become so electronic that one day we won't know how to use a pen. So I hope that even if you have your own personal computer, you'll still keep an old-fashioned journal.

HOW LONG OR HOW MUCH?

Daily! One word--or a thousand, but daily. I don't feel an assigned word count is helpful; I'd rather ask for a specified number of minutes per day. Your teacher will tell you. On the day that you feel so rotten that you simply can't bring yourself to write in the journal, I think it would be great if you opened your notebook and wrote: "I feel too lousy to write today." But if you have many entries of that sort, you need to question your sincerity.

OTHER POINTERS:

- Always **date** your entries.
- Use green, yellow, brown ink, depending on your topic or mood.
- Try dictating aloud to yourself as you make entries.
- Some people carry their journals with them always.

MORE TOPICS:

On some days you may not want or need to discuss the progress of your writing in your journal (you wrote two pages on that subject the day before, say, or things are going smoothly for the moment). On the next page, then, are some ideas for other topics to write about:

Why is my best friend my best friend?

What's my biggest fear today?

My greatest ambition?

The harder I work, the luckier I get.

Dear God, please give me patience--right now!!!

My greatest pleasure is . . . .

Let me draw a map of my soul and plan a trip:

How can I sleep in class without getting caught?

The biggest injustice I know is . . . .

My greatest gratitude is for . . . .

TV is not an art but an appliance.

Those who strike terror in others are afraid.

I'm a little wooden box. These are the things in the box:

Whom do I know who is never, never wrong?

Do I know anybody who is always wrong?

I don't know anybody today. Poor me!

Today I can't help myself but I feel so good because . . . .

Finally,
here's the
most exciting
subject to write
about in the whole
world: Go look in the
mirror and write what you see!

Irina Khlevner / English 090-529 / March 10, 1985

Dear God, Give Me Patience--Right Away!

I´m sitting in my business class, trying to listen to the professor, but I cannot force myself. The class is boring, the professor is a bore. He is either talking in a monotone or making some unfunny joke. What a jerk he is! Although I´m listening to him, I´m not hearing anything.

It´s great that I am not wearing a watch; otherwise I´d be looking at it every other second, and that would annoy the professor. As a matter of fact, I suppose I´m already annoying him now anyhow because I´m writing this down and he knows I don´t take notes when he lectures.

I don´t see the point in taking lecture notes since the first thing Professor Anonymous does when he enters the class is open the textbook that we all have. He gives exactly the same definitions and examples as the text--line by line.

Just now I looked at him for a few moments. So much emptiness in my eyes! I guess he is asking himself: "What is she thinking about? This girl is the dumbest student I ever had." But he is wrong. I´m not dumb; I´m just bored to death by his class. Twice a week I receive punishment for my sins by having to come to this class.

I don´t think God heard me when I prayed for patience--right away! I still have no patience with this man. I think I´ll explode in another minute if he does not stop repeating what I read in the textbook last night. Please, God--hear me and let the professor stop!

Oh thank you, God. You heard me. The professor just said, "The class is over. See you Thursday!" But why didn´t you hear me before? Don´t you understand my suffering? My class mates do. I´m so impatient!

5 / DEFINITION OF TERMS: development vs. editing

You met the term DEVELOPMENT in the last chapter, and we now need to clarify its meaning here because it's a key word in my approach to writing, especially as distinguished from my concept of EDITING. A word of caution before we proceed: not every comp text or English teacher uses these terms in exactly the same way; I claim no universal agreement on these definitions. Yet your grasp of the use of the words **development** and **editing** within the context of this book is crucial.

To develop is to write. It's that simple. "Write about the idea of democracy" is the same as "Develop the idea of democracy" [in writing]. The first time I brought this terminology into one of my comp classes, Thomas, a talkative, conscientious student, asked, "Why don't you just say **write**, then?" This is what happened:

"A good question, Thomas. My response is that so many people have weird notions about what it means to **write.**"

"Like what?" asked Daniela, who was too shy to speak at length but liked to throw out quick questions.

"Some people feel that writing means doing research, checking the spelling, conjugating verbs, spreading white-out all over the pages, and who knows what else--almost anything but the actual composing of draft, the putting down of one word after another on the page or on the screen."

"Aha," said Thomas. "**Composing,** you said. That must be the essence of your message. May I use the word "create" draft perhaps to clarify your meaning, sir?"

"Absolutely. You're on my wave length. Development, within our context, is the creation of a written draft."

"But I don't understand. This is supposed to be a course in plain English composition, not an art course, not a creative-writing class for doing fiction and so forth," said Daniela in an unusually long statement.

"You're right, Daniela, yet all of us have a bit of the artist in us, and the creative imagination is necessary for writing good **non-fiction** as well, which--as you correctly pointed out--is the subject of our course. Moreover, creativity becomes apparent only after you've produced some writing. As a writer, you can't wait for it to appear in your mind. You must go ahead and write, and produce words--in fact, bulk or **quantity**--even before you'll know if there's potential quality."

"You make no sense," said Daniela courageously." I always was told it's **quality,** not quantity that counts."

"Your're right again, but ...."

"Quantity first, then quality," said Thomas.

"Yes. We're getting there. The product of writing is concrete, meaning that it exists out there for everybody to see or touch, not just in somebody's mind. The best way of **developing** your point or idea, then, is to bring your meaning out of yourself, as soon as possible, onto paper or onto the screen, so that you have some language, some quantity, to work with. The next step in the development, of course, is **improving** what you've drafted, the essence of developing your ideas."

Sebastian, whose hobby was photography, said, "When I **develop** a negative, I take an obscure image and make it clear for everybody to see."

"Bravo."

"Too much philosophy," Daniela said. "To the point, please."

"All right. Don't worry about being correct. Think with your hands. Think with your pen. Think with your keyboard."

"Think with your [unprintable]," somebody whispered.

"When you sit down to write an essay, I suggest you actually start to write at once, doing the thinking **as you write along** and as you later go back over it."

"Go back over it?" asked Thomas. "You said you were going to define **revision**--isn't that what 'going back over' is?"

"Precisely: **revision** is **more development.** That's it."

"Wait a minute," said Joachim way over in the corner group. "All this development, all this writing, all this going back over--all this revision means a lot of writing, doesn't it?"

"Yes."

"That's going to be hard on that computer."

"Number one, you can do all your drafting on your yellow pad first, if you wish. Number two, most people find that after they get used to the computer, it speeds up their writing tremendously."

"But even with the old pen, it's hard to do so much writing, so much quantity. I need help just getting started, let alone writing page after page. I'll never be able to write so much."

"You will. I'll help you. See the chapter about **Automatic Writing!"**

Another student had an expression of confusion on his face and almost shouted,

"But if **revision** is 'going back over it,' then what's **editing**?"

"All those things that are not writing that we still have to do before a composition is ready for the readers, ready for publication."

"For example? Grammar?"

"Yes. Fixing your script for correct mechanics and grammar."

"You mean that's not writing?"

"Not in my book."

"You'd better explain yourself and define your terms, sir."

"Let me put something on the board for you to look at":

DEVELOPMENT

You develop a point, idea, or (hypo)thesis with supporting evidence, arguments, examples, stories, specific details. This is the first step, the nucleus of the writing process.

To **develop:** "...to bring, grow, or evolve to a more complete, complex, or clear state." (**American Heritage Dictionary.**)

EDITING

You edit a fully developed draft to prepare if for publication (for readers) by checking it for correct grammar and mechanics according to the dialect of standard English.

This second and final step in the writing process chiefly involves taking out **non-standard** dialect words, forms, and punctuation and inserting standard ones.

"I don't do it that way," said Joanna, who had been silent so far. "I mix it all up. I write and develop my idea for a few lines, and then I edit them, and then I develop some more, and then I edit again, and so forth. You said we're free to choose our own process."

"Are you happy with you own process, Joanna."

"It's never worked for me. Why hasn't it?"

"You asked an excellent question. Here goes: the truth is that for almost all people, the two acts of developing and of editing are not in harmony but in devastating conflict. That's why I put "vs." (versus) in the heading of this discussion. The mental powers that you employ in the two acts are so different that you cannot usually do them both at the same time. One inhibits the other. Let me put this on the board, too; it's a good rule for most writers:

DO NOT START EDITING WHILE YOU'RE STILL DEVELOPING YOUR DRAFT.

"One way of explaining why it is so important to make this distinction between the two stages of making a piece of writing is that the mental faculties that are helpful to your editing ability interfere with and are harmful to your creativity. The editorial voice within you keeps saying to you, 'Watch out--isn't that a mistake, you no good.... And look at that comma....!' This is excellent, of course, if you happen to be in the editing stage, but it is a destructive voice if you are in your creative development stage. So the solution is to keep them apart, like happy neighbors with a sturdy fence."

"You mean I should develop my essay on Wednesday and edit it on Thursday?" said Daniela.

"Good idea."

"But what about an essay test in class?" Thomas asked.

Daniela had the answer again. "You write the essay first and edit it during the last few minutes."

"Sounds good. But understanding the theory of this is not enough. We have to work at changing old habits. You see, for some diabolical reason, most inexperienced writers feel compelled constantly to go back and edit their writing before they've finished developing it. If you have that compulsion, we're going to get rid of it together in this class. It's well worth the effort because superior performance has been reported when writers consider the two tasks as **separate** and do them one at a time. I know that I myself get nervous and distracted if I fall back into that old habit of editing my grammar right in the middle of creating a draft. Do you enjoy cooking and washing dishes simultaneously?"

"I don't like either," said an anonymous voice. "My mother cooks for me."

"OK. But I am not going to do your writing for you, or your editing either. I'm going to train you to become self-starters and self-editors."

"Self-editor, you say?" said Joanna. "I've never been able to find my own mistakes. In a work book, yes. In my own essays--forget it!"

"I'll show you. See the chapter about EDITING. But later for that. In fact, to help you form the new habit of separating development from editing, I want you to forget all about editing--that means no grammar--for the first few weeks of this course. In the beginning we'll cultivate development only."

Thomas now had a wise look on his face, raised his hand and said, "The lesson you've taught us today is called 'the rhetorical mode of definition,' right sir?"

"Yes, in one sense we could say this has been an 'extended definition by example.' But I prefer to say that we have **explained** some concepts useful to this course. And the rhetorical mode that **explains** usually is called **exposition.**

"I feel cheated," said Sebastian. "All this time I was learning about definition."

"How do you define **cheat,** Sebastian?"

"That's easy." He fumbled with his pocket dictionary. "Here: 'to deceive by trickery; to swindle,' it says."

"Good. That's called **definition by synonym.** Another type of quick definition is through elimination or negation: you say what the word you want to define is **not.** For example, earlier today I said that development is **not** editing for grammatical correctness. There's one more type of definition...."

"In eleventh grade," said Joanna, "I studied **logical** definition I think they called it. I don't remember what it was."

"Yes, by classification. That's a formal way of defining words and concepts. This is how you do it. Let's say you want to define the words **democracy** and **dictatorship.**

1) First you put them in the **class** of "form of government." You can see that they both fit in the same class.

2) Next you distinguish the two words by presenting **features that separate** them from each other within that class:

a) **democracy:** leaders are freely elected;

b) **dictatorship:** leader is self-appointed; holds power through violence.

"Is this class a dictatorship or a democracy, sir?"

"That, Joachim, is a red herring, or a false issue. And these are terms I hope to discuss with you when we get to **argumentation** later in the course."

Joanna was wiggling her foot and looked impatient. "If this is a writing course," she said, "why are we talking so much? When are we going to write?"

"Write now. Turn to the chapter on **automatic writing.**

ooOOOOOoo

There's one more lesson in this chapter. Beside the definitions and explanations of terms like definition, explanation and, above all, **development** and **editing,** I have also offered a demonstration of the use of DIALOGUE. Let's look at the **form** first:

° No more than one speaker per paragraph.

° The quotation marks that go before and after the spoken words also go **outside** commas, periods, and questions marks.

° If it's clear from the context who the speaker is, you can leave out the "I said" or "she asked" phrases.

At this point you might wonder why I'm pushing dialogue. Isn't essay writing really monologue? Yes, most academic or business writing is in monologue form, so to speak. Yet here are some reasons why your experimenting with dialogue will be of benefit to you--and your readers:

° You'll find it easy and enjoyable to write dialogue. This means that you'll write **more** (quantity). The reason it's easy is that it's close to speech. In fact, human speech probably originated as dialogue. Have you ever noticed that a baby's "monologue" actually is a dialogue?

° Some of your writing assignments will be based on interviews. Parts of your notes--your raw material--could be retained as dialogue in your finished essay.

° Most people love to come across dialogue in their readings; it enlivens the prose.

° Supporting evidence in the form of direct quotation from a knowledgable source is often convincing.

If you get really excited about writing dialogue, later in the course you'll want to see the chapter with the one-act play in the back of the book.

6 / AUTOMATIC WRITING [freewriting]

So it's time to start this chapter shouldn't be too hard though really it's too early in the morning for anything, except ~~to~~ writing. So I don't need my pen now let's go ~~to~~ to the machine

[Errors have been retained to give the flavor of the original.]

here I go again I really feel better at the computer than with my ugly pen I mean handwriting. Automatic writing. Used to call it free-writing. Same thing. I'm doing frags now, so what. I'm a free man, this is a free country, right, I'm not a slave. I'm a FREEWRITER!

Really got to get busy now. The pount yes what is the point? Too early Glorious sun almost blinding me in its reflection in the window accross the street, spotted dog playing on the roof, do you spell across with one or two c's??? Whatever you call it, this automatic stuff is a warm-up that cleans out the garbage in the front of the brain so I can get to the real stuff, so I can get focussed: muscles in fingers and arm, heart, heart of the matter mind or heart ? What's ***sponteneity,*** *honey? Honesty? Honestly? Which comes first* ***chicken or egg,*** *mind or muscle, hand or brain?*

°oOOOo'oOOOooOOOoo°°. ha--this is more fun than writing!

Oh, here's an idea: writing is writing, and nothing else. Training myself to the action of doing it, namely writing instead of other things, will get me into the process so the writing will get done, done, done. Catch a falling star, said John Donne. I learn to hold on to the falling words out of my head before they sink in the ocean--&$Δ%*&²+=¶Ω?/@!#$%&$()ΔΔ%%$($

There, fore, ladies and gentlemen, it's crucial at this moment in history to realize that there's no substitute for actual writing. grammar rules are no substitute. Vocabulary building is no substitute for writing. Research may help but it's not writing. Thinking is fun, but it's not writing. Praying . . . well, while it's really not writing, it may help in certain hard cases....But writing as muscular, that's the real McCoy. Both that and it gets to the secret heart. Both. Back and forth, crazy now I'm ready. What time is it oh my, it's ***up!***

That strange writing on the previous page was a piece of my own automatic writing that I did just prior to writing the first draft for this chapter. As you could see, I started out with my pen, but because by now I'm so accustomed to the computer, I quickly moved to my machine and fired it up. I included those scribblings here for three reasons:

1. to give you a sampling of what automatic writing might look like.

2. to impress upon you that automatic writing is not for beginners only; it can be valuable for for writers at all levels of experience.

3. to show you a special benefit: when you're lucky (automatic writers are often lucky!), your exercise will not only warm up your arms and brains, but may even produce some writing that you can use in a current project.

Let me give you an example of number 3. Once I got warmed up in the sample I just gave you, I managed to hit upon what I think is the crux of the nature and value of automatic writing: it does **two jobs simultaneously.** Let's see what they are:

° Automatic writing trains you to perform the mechanical engineering act of transferring brain energy to the screen or page.

° With experience, your automatic writing will help you release subconscious insight. It will tap your heart, if you wish.

The first point ought to be important to you--unless you already write fluently in great quantity every day--because although writing seems to be "like thinking" (and does **include** thinking)--it's actually a discrete, specialized activity. Your **thoughts,** no matter how brilliant when playing around in your head, are not writing **until** you translate them through your neck, shoulders, and arm(s) onto the page or through the keyboard onto the screen.

This part of the writing process has often been ignored in the past. Although writing conveys **meaning** from your mind, the channeling involves muscles and nerves of your body. They need practice until they become nearly automatic like reflexes. Surprising as it may sound, then, to some extent writing is a muscular activity. In the beginning of my writing courses, my students complain that their arms get sore from "too much" writing. That's good. The soreness that comes from exercising always goes away, as you know.

Now let's look at the second point listed above. Your own mind is a wonderful storehouse whose riches often are not brought into daylight. But the very act of writing affects your thinking by revealing ideas you were not aware of having **until** you saw them on the page or screen.

You can see then, how foolish is this rule, often given to students: "Don't start writing until you know exactly what you want to say." I'd call that dangerous advice and hope you'll never follow it. In reality you won't know exactly what you want to write next **until** you have begun to write and you can see your word and ideas unfolding right in front of you. It can be a magical experience full of new insights; you probably won't believe it until you see it. To see it you have to do it--and the best training is automatic writing.

"I've tried freewriting or what you call automatic writing once and it was a mess," said Joanna.

It's supposed to be a mess. A great writing teacher, Ann Berthoff, talks about the necessity for having **chaos:** the writer is challenged to make **meaning** by moving from chaos to **order.** I firmly believe that the most common cause of writer's anxiety is the crazy, perfectionist effort to bypass chaos or that messiness that Joanna brought up: to try to start out with order is an invitation to paralysis and permanent chaos.

The writers who persist in trying to write a final draft at once almost always end up sounding unoriginal and boring in their essays. They miss out on all the wonderful **discoveries** the rest of us make with our messy first drafts. And the best technique for starting a happily chaotic draft is automatic writing.

You can do it with pen and paper or with your computer, along with your class, in your group only, or alone. It's best to get used to automatic writing with others; there's something stimulating about sitting in a room quietly writing with a group of other people. The communality is reinforcing (often the teacher writes along with you). Of course, once you've become comfortable with the process, you'll benefit greatly from doing it alone. But start out in your group or class:

° Get your pen & pad or computer ready.

° Stretch your arms, take a deep breath or two, close your eyes for a moment, and tell your body and mind to **relax.**

° When your teacher or group chair says, "Go!" start covering your page or screen with words.

° Write as speedily as you can, yet without rushing, but above all NEVER STOP--until time is up. It's **non-stop** writing. If you get stuck, write "I'm stuck, I'm stuck" or whatever, until you get moving again. It never fails.

° Remember that you're FREE to write anything and that **nobody is going to criticize you.**

° Spread your words all over. Don't talk, smoke--or stop for anything.

Usually these sessions are timed to last 10 minutes. You'll often find, though, that when the time is called you don't want to stop. You're getting excited; that's a good sign. But do stop and look over what you wrote. You might be shocked at how much you wrote in just a few minutes (sometimes more than a page) and how good it looks. You may want to mark the best parts in the margin. Perhaps you can use them in some other writing project.

One view of automatic writing is that one should never share it with others. A more liberal approach, which I embrace, is to allow writers the option of reading or showing their it to their group mates, class mates, or friends if and when they have a desire to do so. If somebody should show you a piece of automatic writing, whatever you do, please don't criticize it. If you can't say something nice, just nod, smile, and return it. (More about feedback is coming up in a later chapter.)

Before we move on to talk about some specialized types of freewriting on computers, let's look at a couple of samples:

> *On avoidance of writing. I must have this notebook only, or this pen. This is the wrong pen. I can't write with this. I must have a new notebook. The pen doesn't feel right. The lighting is bad. I have to do something else foist. I must clean my room first. I must write that check first. I must get dressed first. I must take my shower. I must have a cup of coffee first. I must eat breakfast first. I must digfest my breakfst. I must wash the dishes. I must get my laundry together. I must lie down and pull the cover over me.*
>
> *I must check my money. Do I have quarters for the bus? Do I have tokens for the train? Did I finish my book? Where is my book? go to the bathroom? Do I need a glass of milk? Is my hair brushed? Is my watch on my wrist? Did I brush my teeth today? What's on the radio? What's Mom doing?*
>
> *The phone is not ringing.*
>
> *Maxine Berman*

If you compare Maxine's piece to the one I wrote at the beginning of this chapter, you can see that hers--though not finished--is much more coherent and structured than mine. Why is that? It's because we're all different and write with different styles and rhythms; she writes in her way and I in mine. What's **your** way and style? Do you want to see a really messy piece?

oh no not again can't stand this writing writing I am stuck period what no I am going to try. No my mother this what am I doing what is this topic it stinks I know it's write about really right that I have senses. Write about your senses but I have no senses I am not sensible they all said so I am very sensitive and now I am stuck very stuck when is this going to end.

I have 5 sense the cat has 9 lives shut up when is all this automatic bull going to the dogs, raining outside stuck. I'll write tomorrow. Tell mother about the dog tomorrow.

[Anonymous]

This is a fairly typical--and successful--example of automatic writing by a beginner. I called it successful because as long as you write **something,** you have succeeded in the exercise. Remember, all you have to do is cover the page or screen with words, any words.

If you haven't had one of these sessions yet, let's do one with pen and paper now. Are you ready? We'll do a short one. Ask your teacher or your group chair to time you for five minutes. First glance again at the guidelines I gave you two pages ago. Then get out your yellow pad. There'll be a choice of two topics, one basic and fairly broad one, and one more specialized:

1. "Sensory Perception: the Here and the Now."

Look into your body and mind and write about what you **see, hear, feel** and **smell.** Can you hear your heart beat? Even if everbody in the room is quiet, there still must be some noises. Any traffic or other outside noises? Describe them! Is your body tense or relaxed? Are your palms dry or wet? Is your mind empty or bursting? What do you see in the room? Bulletin board? What's the teacher doing? What's the taste in your mouth? Get all the details down on the page. (If you feel like writing about something else--go ahead. The main aim is to cover the page with your words.) Go:

2. Write about a **significant writing experience** you have had, either recently or some years ago. Here are some examples: "My Worst English Teacher"; "My Best English Teacher"; "The Time They All Laughed at My Writing"; "The Day My Writing Was Praised"; or anything similar. Go:

ooOOOOOoo

I recommend that you do automatic writing every day that your group meets. Later you should try it at home, too. (See the chapter on your log or journal.) It will have a liberating, enriching influence on your writing career.

INVISIBLE AUTOMATIC COMPUTER-WRITING:

Now let's look at freewriting on the computer. The first thing to note is that you can write non-stop, relatively fast on the machine pretty much the same way you do it on your yellow pad. In most cases, however, your automatic writing in the lab will be either alone or with one or two neighboring group mates; it's difficult to get all the machines coordinated in the lab. But we need to deal with a minor difficulty that some people get into when they freewrite on a computer.

What occurs in some cases is that we become error-anxious when hammering away fast on the computer. Because of the clarity of the text on the screen, we get annoyed by the typing mistakes that stare obnoxiously down into our faces. And what happens then? We stop and go back and EDIT by correcting the typos. Of course we're not supposed to do that, especially during automatic writing, which is development, as we know the term.

Some of us learn to resist the temptation quickly and to go free of the editing mania. But for those of us who fall into the correction-trap, there's a remedy: "invisible writing." It's an ingenious method for writing on the computer developed by Stephen Marcus of the University of California. It not only stops us from editing prematurely, it might give us a creativity boost. While some people don't care for it--some swear by it. You have to decide for yourself. Here's my simplified modification of the invisible-writing method:

Adjust the brightness knob on your monitor so the screen goes blank. Then time yourself and have a vigorous go at the keyboard. Since you can't see what you write, you will not be distracted by any mistakes, and you will be able to focus on the train of your thought. Later you turn the light back on to reveal your writing.

Let me tell you what I just did: I turned off the brightness on the computer I'm using now and went ahead and typed the paragraph you're just now reading (corrected for typos later, of course, since I'm not an expert typist). When my phone rang just as I started writing, I chose not to answer. This fascinates me. I seem compulsively focused on my train of thought. I have to tell you that it is a good feeling. I tried it once before and wasn't convinced that I benefited from this method, feeling I could take it or leave it. This time around I feel better, and I may be changing my mind. I am concentrating sharply. So I urge you not to give up on it the first time you try it. Tell your group mates and your teacher how you feel about "invisible writing"--share your experience. Now I am going to turn the light back on.

DIALOGUE ON THE COMPUTER:

You need only one machine for this enjoyable exercise in which you **pair up** with one of your group mates. Sitting close together in front of the keyboard, you take turns in asking and answering questions in a written **dialogue** that shows up on the screen and later on in the print-outs (get a copy each!). It's interviewing, in other words. This is perfect for warming up and, more specifically, for the assignments in the chapter called **YOUR GROUP**, coming up soon. See also Chapt. 9.

YOUR VOICE:

We've already stressed how helpful automatic writing is in getting us started writing and in creating quantity. Another benefit is its character-molding force. Practicing freewriting vigorously will help you create your own personal **style**--your **voice** as a writer.

When we read business reports or scientific papers, we often see an impersonal, objective (boring) style. Yet most types of nonfiction writing, including essays written in school, affect their readers more powerfully if presented from a fearlessly **subjective** viewpoint. That means writing in **your own voice** and coming clean about being **"I"** and not "you" or "one" and not using the impersonal, passive **"...it is felt that...."**

Effective writers don't beat around the bush about how they feel about their subject, their readers, and themselves. I'm convinced that **honestly emotional writing** is strong writing, and that if you want your readers to stay with you--your primary purpose--you must give them the real stuff, which is a piece of yourself. Throw in some tears and laughter, too, if you can. They will turn your pages and ask for more.

"My last English teacher told me the opposite," I can hear one of you saying. "I was trained never to write "I" and always to be objective and formal."

Yes, I know. I was taught the old-fashioned way, too. But we want to change along with the changes out there in the market place of our society. Compare the following two statements, both saying the same things--but with different voices:

"We are endeavoring to construct a more inclusive society."	"We're going to make a country in which no one is left out."

An American president made one of those statements, and his speech writer (a pro) wrote the other one as a revision of the president's original. Whether we like it or not, the **voice of our writing** reveals and projects our personality, thus affecting the way our message is received. Which sings best in your ears--the statement on the left or on the right? (Discuss in your group the reasons for your choice.)

Although it's a personal matter what level of formality we prefer, there's a definite, strong trend modern American writing--academic, business, and general--toward **relaxed informality.** Even the banks and the federal government are now revising their forms and documents for greater simplicity and readability. That old official style goes under different names: officialese, bureaucratese, gobbledygook, and Engfish, another term for **studentese** which emerges when students imitate the dying, official style.

Your group may want to discuss **why** there's such a thing as officalese, in English (as well as in other languagess) and why more and more writers now shun it.

Assuming that you desire to find your own personal **voice,** how do you go about it?

If you're concerned that you might not have a writing voice in you, I sympathize with that thought. But we all do have a real voice inside that's trying to come out. And since I have chosen to talk about it in this chapter, I clearly feel the technique that will bring your own style out into the open is AUTOMATIC WRITING--and **more** automatic writing. It has been proven a winning technique. In addition, focusing on YOUR AUDIENCE (another chapter coming up soon) is very helpful in learning to modulate your voice.

Some people write too much.

Intermission -- Sample Student Voice:

NAME OF GROUP: Untouchable Writers

Date: 10/10/87

GROUP REPORT

Everyone is very busy in the lab. Now most of us, basically, are mastering most of the computer functions. There's no time to talk, no time for procrastination. We are all "busy bees" these days.

We finished up in the lab and proceeded to the classroom on the 7th fl., N781. On the way a class mate approached me about the stockmarket going down. I voiced my opinion and answered his question, "Is it a good time to buy stocks, while the Dow Jones is down?"

I answered him, "Yes," and gave him a detailed opinion and I also said to him, "But once you make big money from the market, "DON'T GET GREEDY--GET OUT!" Little did I know, for I was so busy writing my "GROUP" essay, that the stockmarket crash was going to be the topic for our 10-minute Essay in class tonight.

The professor is playing tough coach today in the classroom because he is upset at several people's journals. I'm not going to name any of my group mates who are in this category. The professor said he loves us all--before implementing a process of scolding those students who neglectfully don't write in their journals. This can cause their destruction. The professor then continued with his scolding, then partially relented and stated, "Writing develops a better writer. Saying you can't find the time to write, is like saying you can't find the time to be SUCCESSFUL!"

He then proceeded to the "10 min. Essay" and described its procedure. We then began. I wrote about the a"Stock Market," of course. So did Omaria. Shamar and Mike wrote about the case of Bernard Goetz.

We had a little discrepancy concerning our "group" essay. Shamar, the Group Chair said, "We have to write the same group essay. All of US!" We don't have to do that. Thank God, because I'm just about finished with mine. If I only could type faster, it would be completed already.

I hear, as I'm writing this "Group Report," my colleagues saying among themselves, "That's his first Group Report and he's writing a book!" and I say, "If you're going to DO IT, DO IT RIGHT!" The End.

signed: WILLIAM PURCELL, Jr.

Intermission--Sample Student Voice:

Mary Ingram / English 090-522 / Nov. 12, 1985

Daily Group Report

Today feels good since we had yesterday off, Veteran's Day. Margaret and I are the first of our group members to enter the class. For a while it looked as if there would be just the two of us. Finally Jacob showed up, and Sangra came late. Carlos is absent today, for the first time. I hope his home wasn't hit by the planes that crashed Sunday over in New Jersey where he lives.

The class then did the usual 10-min. essay. Then we got back our last practice-test essay, and we went over the Bulletin together and some other business. By then it was time to go to the computer lab.

In the lab the members of our group, the Pen Pushers, were still in a discussion of the practice writing-tests that were returned today. But we finally put them aside and began to work on our essays on the computer. By the way, I forgot to mention that we elected Sangra our new chair today, for the next three weeks. Perhaps overwhelmed by her election, Sangra is undecided about how to use the computer today, so while she's gathering her thoughts, Margaret, Jacob and I start working on the screen.

We have a visitor in the lab today, an English teacher, if I'm not mistaken. I believe our teacher gave us his name, but I was so involved with the machine I forgot to put his name down. He's here to learn about teaching with computers.

Well, Sangra has collected herself and has started on her computer. In the meantime Jacob and Margaret, the Whiz Kids among the Pen Pushers, also known as the Computer Kids, are very very busy typing away. The lab is a lively hubbub of noise. At one point Dr. Moberg comes over to our group to see what we're working on. He complains that we're not asking for his help any more. Jacob then responds that we're now able to help each other.

I believe I was the first member to make a print-out today, which is good for me, because I'm usually the last to print. Then the Computer Kids came next. Sangra, realizing that she hadn't put her disk in the machine, is horrified and thinks all her work is lost. But all is not lost. All she has to do is SAVE. And off to the printer she goes.

By this time the lab session is just about over. The Pen Pushers had a very productive lab day. You know, the computer lab is becoming a nice place to visit, instead of a scary one.

signed MARY INGRAM
GROUP REPORTER

7 / GROUP REPORTS

Although your group gradually will develop certain initiative, independence, and authority over its own affairs, the suggested structure of this course imposes two minimum organizational requirements, as mentioned in chapter 1:

> a) CHAIR (short for chairperson), who leads group discussion and usually serves for two or three weeks. Everybody is chair at least once a semester. (In my classes, the chair also hands me a daily attendance sheet, but that would be up to your teacher.) Most importantly, the chair keeps a roster that determines who is the day's group reporter and makes certain that a report is being written for each class session.
>
> b) GROUP REPORTER, who writes a description of what happens in your group, both in the classroom and in the computer lab.

The purpose of these reports is not chiefly to permit the teacher to keep a tab on what's going on in the various groups--though that is one benefit--but to give occasion for further writing by at least one member of the group. The reports also add seriousness of purpose to the proceedings: a record is being kept.

So the **group report** serves many useful functions, and doing this assignment with care and enthusiasm will benefit you, your group, your class, and your teacher. To define the term "function" here, let me give a couple of near-synonyms: **justification** and **role**. The assignment to write a report for each session is not just "busy-work" invented to make your life hard but is justified by several helpful roles:

1) Because of your decentralized classroom (groups often functioning as separate workshops), your teacher doesn't always know the exact character of the work in your group. Hence your group report offers a record of information of the highlights as well as some of the details of your session. Your teacher can coach you better thanks to this information.

2) To an even greater extent, the members of the other groups don't know what's happening in your group while they're busy writing and giving feedback in their own groups. Perhaps they could learn something from your experience. Thus sometimes--when a well-written report offers pointers useful to the whole class--your teacher can make copies (with permission of the originating group, of course) and distribute them for general class discussion. Even better, on occasion your group could take initiative and show a particularly good report to the others either by making copies or simply putting the original on the bulletin board.

Writing on Computers

GROUP REPORT

Name of Group: ______________________ Date of Session: __________

REPORT:

Please describe the writing and talking of your group in today's session--both in the classroom and the computer lab. Support your general points with detailed examples. Who said and did what?

[suitable for duplication]

signed ---------------------------

Group Reporter

[Use the reverse if your wish!]

3) When **you** are your group's reporter, you **do** know what's going on, but you'd **learn** more about language communication if you have to write about what you see and hear. You'll sharpen your **listening** skills, in other words. And did you know that the very **act of writing enhances learning?**

4) GENERAL AND SPECIFIC: Summarizing what takes place in a one or two-hour learning session is both a challenging and rewarding **writing task.** It forces you to choose and make one or more **general** points, backed up by relevant **examples** with **specific** details. For instance, if you are the reporter and you claim that your group had a "rewarding but demanding" session, you need to supply the specifics that illustrate that general statement. In other words, your task is a **model** of typical essay writing, as I described it in chapter 4. But what's **not** typical is the immediacy of your material, which is a great blessing for a writer in training (see 5 just below).

5) THE HERE AND THE NOW: If you are like many students, you may have felt in the past that much work in English classes seems to call for writing about subjects that seem distant to you, far-fetched topics about which you have no knowledge. And you feel uninspired. I remember many such essay topics from my own high school days: "Describe the Concept of Being a Gentleman," was one. I didn't have the faintest idea what to write about, neither generally nor specifically. But when you report on what's unfolding in front of your face, you have a gold mine as a source for your writing. It's excellent practice.

FORM OF GROUP REPORT:

Paper: Your teacher will either hand out a special report **form** (see illustration on opposite page) or tell you to use your own paper or computer paper.

Length: Again, your teacher will instruct you. In my classes I usually ask for half a page to a page the first few weeks. Later I expect at least a page and a half and give special credit for long group reports.

Pen or Print: In the beginning of the term I gladly accept reports penned in longhand, but later I expect them to be done on the computer and printed out.

When Due: Normally these reports are due immediately at the end of the class. (Occasionally, however, students write long, exciting reports they consider unfinished, so they want to treat them like rough drafts that they wish to revise and edit for submission later as a regular essay assignment.)

Who? What? When? Where?

Why?

...and How?

WHAT GOES IN THE REPORT:

Imagine that one of your group mates is absent. The next time he or she asks you, "What did you do last time?" You offer your group report and say, "Here--it's all there!" Of course it's not "all" there, but the highlights are: whose essays you all read and discussed; who was silent; who said what; any new insights; how you felt about the teacher's visiting or not visiting your group; whatever telling details of excitement, resentment, or humor you observed; and even your own daydreams, frustrations, and joys as a group reporter.

Most teachers appreciate complete honesty in their students' writing. If your group is having an unproductive day, for example, it's best to put your thoughts and feelings about that in the report rather than faking it in order not to "stir up trouble." Dishonest writing seldom succeeds.

What do you do as a reporter on that quiet day when your mates are mostly writing by themselves? Interview them! Say, "Excuse me, Hector. I need some material for my group report. How's your writing going?" Then you get something to write about. Even if the answer is, "Don't bother me--I'm busy," you have **that** to include in your report.

Finally a word about the writer's **attitude.** If you approach your report-writing as you grope your way to the sink to do the dishes, you'll have a difficult time enjoying your creativity. It would be better if you could think of your report not as a chore as much as a mission of honor.

For your report is the real thing. It's **real writing** about the flesh, blood, tears, laughter, and soul of real people--your peers in your group. They will be just as eager to read what you have written about them as your teacher will. In fact, you'd do well to pause for a moment next time it's your turn, and decide for **whom** you're writing your report: your group mates, your class mates, or your teacher? Is there any other conceivable audience? It's **your** choice.

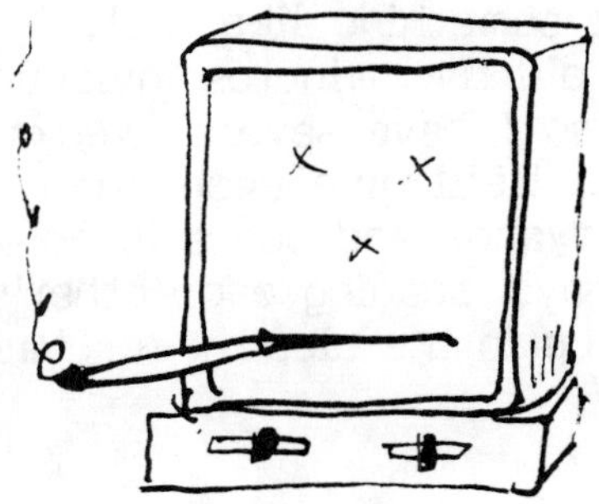

Intermission--Sample Group Report:

Gilbert Johnson / English 201-715 / Feb. 20, 1986

Daily Group Report

At first I was surprised at the topic for our ten-min. essay today: "Can One Be Addicted to Music as to Drugs?" So I was slow starting. Then the more I wrote, the deeper my thoughts penetrated. Yet hit no bottom. I looked around in my group. Gina, Maryanne, Joe, and Pedro in our Critical Writers Groups all were writing, their pens on fire. Nobody in the class wanted to finish when time was up. Never seen it like this.

And so many hands volunteering to read to the class. (I'm probably writing fragments but right now I'm making sure I write no run-ons.) I believe the instructor thinks music can be an addiction, but as he has said himself, that's neither here nor there. All but one of the students who read out aloud, however, were almost angry with the implied position raised by the topic. George in the Flower Group, for instance, wrote that he felt the issue is unnecessary and is a veiled attack on young people's playing rock on their radios. Many said, "Yeah," or nodded in agreement.

That's when the lone, soft voice of Sandra from the Hungry Pens asked to read hers. She suggested a yes to the question in the topic by referring to Baldwin's story, "Sonny's Blues." She asserted, if I could hear her right, that "Sonny's a dual addiction, both to drugs and, yes, to music. That's a short story about compulsion" or words like that. It got a little quiet in the room. The most fascinating part was the discussion that then started up about if you can have a "good" addiction. Joe said he wanted to talk about an article he'd read about sexual addiction. It was about that time the instructor said we could continue talking and writing about addictions later on in the computer lab. It was time now for our daily practice in editing aloud.

I hate that. Both the interruption of our intellectual debate and looking for bad grammar. But I do need it. So I'm regretfully grateful. I only wish my grammar medicine would taste like chocolate. Today I noticed that I read my writing a little louder than usual. Nobody told me to shut up. And I wasn't bothered by anyone else's voice. I caught several fragments. I can hear how I don't want a pause when I have a period sometimes. I wonder why my teachers in the past never told me to edit aloud. I find more and more mistakes every time I try it.

In the computer lab I decided to write some more about addiction and to make a whole essay out of it. I'm going to read Baldwin's story. The more I think about it, Sandra has a point. My title: "Addiction Is Not All Drugs" or something like that. I turned my brightness knob all the way down on my IBM and did five minutes invisible writing on my topic. I will use this as my free-topic essay, so I have several weeks. Gina also very definitely is going to write on the same topic. We'll give each other ideas on the phone in the coming days. Pedro says maybe. Maryanne and Joe said no--they were giving each other feedback on their "interview" essays, reading each other's screens. It was a good day. But I wish we had the computers up in the classroom. Time to print this Report now.

Intermission -- Sample Student Voice:

Name of GROUP: COURAGEOUS WRITERS

DATE: 11/30/87

GROUP REPORT

The Semester is coming to an end. I am partially happy and partially sad. I am happy because I will get to go to bed a little earlier and have more time for myself and I am sad because I may never see my class mates whom I have grown to like and appreciate, especially Pam.

In the computer lab it was very quiet today, at least in the beginning. Michael completed essay #4 PTREV-C. Lloyd completed essay #3 PTREV-B and printed it. Pam completed and printed essay #4 PTREV-C and showed it to the professor. I finished the Emerson essay and printed it. I, too, showed it to my professor and he liked it. He had a few suggestions and corrections.

In the class room our professor talked to us for a few minutes then he turned us into our group-mode. Our chair person, Pam, timed us for ten minutes while we did our automatic writing. The topic was difficult so our writing was not very informative or did not make much sense. I guess it is because we had little knowledge about the topic. "In the news today, we hear a lot about human heart transplants, laboratory-grown bacteria, test-tube babies and other ways of artificially prolonging human life. Many people have argued that these technological advances are unnatural and violate the laws of both God and man." We had to agree or disagree and support our opinions.

After the essay we did different things . We talked about the upcoming exams and what we hoping to get, a "12" of course. Lloyd shared one of his essays with Pam who did some editing. He then showed it to the teacher who gave me his personal evaluation. He offered to make copies of one of our essays to be exchanged with our class mates for feedback. I gave one copy to William and one copy to Pam. Class continued like that with plenty of feedback, both in talking and writing until 7:35 when class was over.

signed: CHITREKA SINGH

Intermission -- Sample Student Voice:

Miguel A. Reyes / English 090-084 / Nov. 3, 1988 / #5

THE BREAKFAST CLUB: LOVE AND MARRIAGE

What's going on? What are we doing? Is everyone doing their work? Is anyone having trouble? How about my morning kiss? Did you have breakfast? Are you still half asleep? What's bothering you? Why are you being so snotty? I can't deal with you anymore!

It all boils down to marriage. Let's see if our group can imagine our married state of life. Well there are Miriam, Naranja, Gertrude, and Nancy. These are the women who are members of the Breakfast Club. Keith and I are the only two men. To think how lucky we are--being surrounded by all these women! Or are we really lucky? Let me give you my point of view.

I'm actually married to each and every single member of the Breakfast Club. It's funny how in our group we have to actually love each other, or just hide it. Maybe there are two or three members of the Breakfast Club who would like a divorce. The bad thing about it is that we're all actually married to each other. We can't do anything about it. Or maybe we can?

Let me now look into a real marriage. Maybe there's a solution to my marital problems. In a real marriage there's love. There's understanding, cooperation, and even a sacred bond of friendship. The Breakfast Club has neither of those qualities to bless our married state. Maybe this judgment is a little too harsh. It might not include all the members of our group. There might be one or two members who feel differently about another member of the Breakfast Club. In any case I wouldn't want word to get out. So I'm keeping this very confidential. [Names have been changed to protect the innocent.]

In any case, if I want the Breakfast Club to become a successful group I'd rather get back to my fantasy. The fantasy of me being married to the members of the Breakfast Club. Miriam, Naranja, Keith, Gertrude, Nancy and me Miguel are still married and yet still looking for that love. The love we need to set us aside from all the rest.

These few words I have just written make up my idea of love and marriage in our group. I now want to focus on the Breakfast Club's individual ideas of love and marriage. Let me move on to the first member of our group. I'll begin with Miriam, a very friendly and talkative, high spirited young woman. She has her own feelings of love and marriage. She feels love is very strong, but there are always risks involved, such as as getting hurt. She feels that when you get married, you're committing yourself to that special someone. She stated that she's in love now, and in love with someone she's not with. I hope she get's everything clear because getting hurt from love is the worst thing that can ever happen.

Naranja feels that love is fun. She states that she in love at this very moment. She's been in love for exactly one month, two weeks, and nine hours. This was recorded at 9:16 A.M., on October 17, 1988 the day of her interview. She says "Marriage isn't good unless you are ready and you feel that you're with the right person." I agree with her one hundred percent.

Keith, the other man in the group, has a strong point of view on the subject. He states that love is blind, and it's certainly dangerous. He suggests that you should disregard people's advice. He thinks that you should go with your emotions. Marriage to Keith is a very big step. He suggests to everyone that people should make sure that they commit themselves to that right someone. I believe that Keith wants everyone to watch out. "Remember," as he said, "love is blind."

Gertrude, the young lady with the beautiful long hair, is next. She feels that love is a good feeling. When you are wrong, you could be hurt very much. If you decide to be married, you had better be sure you commit yourself to that person. Gertrude is definitely going to watch out for herself! I think that Gertrude is alking from experience. I'm going to remember those few words of hers.

Nancy, the quietest member of the group, has participated in the subject of love and marriage. Nancy distinctly feels that she has been in love and that she was hurt. At this moment she's not in love. She also never really thought of marriage. I think that Nancy has been hurt but doesn't really want to talk about this subject. I respect her for her thoughts; she has contributed richly to the group.

It brings me great pleasure to have introduced the members of the Breakfast Club for the last time: Miriam, Naranja, Keith, Gertrude, Nancy, and me, Miguel. I hope that we'll always remember that Love and Marriage are not always what they appear to be!

8 / MEANING: GENERAL AND SPECIFIC

Have you ever heard about the fox trying to get to the grapes? He was wandering through a luscious orchard one hot summer day when his sharp eyes caught the sight of a bunch of juicy-looking grapes hanging from a branch. "Just what I need," the fox said to himself. He stepped back a couple of paces and jumped up to catch the fruit with his jaws but missed them. He tried again, this time reaching with his paw, but in vain. He concentrated hard and jumped again with all his might, but he failed. After he missed again, he felt so exhausted he gave up. "They're sour," he then said to anyone who might be listening as he trotted off, nose high in the air. "Sour grapes," the fox repeated.

Like the story about the inventive, thirsty bird, this is an old Greek fable by Aesop. You probably know many others--about crocodiles, lions, lambs, and so forth. And then we have all the super-short tales, often about animals, called **proverbs,** such as "You can lead a horse to water but you can't make him drink." I want to raise a question, however: are all these stories about animals? Or are they really about human beings?

What's the meaning of all this, anyway?

Isn't it childish for us here in an academic English course to talk about animal stories? Not necessarily. Let's go back to the fox fable above: have you ever heard an unmarried person say that marriage is no good (sour)? Or to pick something more immediate, do you know anyone who wouldn't write on a computer because "computers are impersonal"? The truth is that old fables, fairy tales, and proverbs usually contain rich meaning that can illuminate your own concerns today. But now to the point of this chapter.

I've met many students in my teaching career who thought that an English course is about grammar and avoiding errors. But it isn't (except in a small way). It's about **communicating meaning.** Here's what Bruno Bettelheim, an author and psychologist wrote:

> If we hope to live not just from moment to moment, but in true consciousness of our existence, then our greatest need and most difficult achievement is to find the meaning in our lives.

What's going on here? We're going from kiddie stories about animals to the meaning of life? Rather fancy development. Well, **development** is a key term (see the next chapter, too). You see, writing is not only a technology (pen, paper, computer, printer), but an expression of your mind, your meaning.

"What do you mean?"

What I mean is that going from the "small" (specific) to the "big" or "fancy" (general) idea is the way we humans normally think. Understanding this bouncing back and forth between generalizations and particulars helps us when we write.

Actually, what I'm saying is not as new or different to some of you as it may sound. It's really an old principle from your earlier English classes: you support your thesis (general) with examples and details (specifics). If you begin with the general idea, it's called deductive; if you begin with the specifics, it's called inductive--right? The important thing here is not the terminology but the assertion in the next paragraph.

The general point I'm making now is that in reality our mind doesn't travel that straight a path. Instead it usually bounces back and forth between the big picture and its detailed parts. For instance, reading about the thirsty bird's efforts might make you remember a whole series of things, something like this: a) your own trying to learn a **specific**, difficult idea in grade school; b) a **general** feeling that life can be difficult; c) **specific** birds you've watched; d) sweat in your palms that time you solved a problem (specific); and e) a reassuring **general** idea that perseverance pays off.

In the exercise below, make a similar list about how your mind might deal with the issues raised in the fox fable:

a)
----

b)
----

c)
----

d)

Obviously the finished writing you present to your readers ought not to include all these wanderings of your mind. But **your awareness** of them will help you reach the clarity and organization that both you and your readers desire. If you try to bypass the wonderfully rich process of your mind, you might miss out on much of the meaning it contains.

Your work will bear sweeter fruit if you look at the bouncing-back-and-forth pattern of your thinking that I mentioned. In the exercise about the fox, did you indicate which item was general and which specific? Once you see and accept what your imagination is doing, you can consciously create order out of what may at first appear chaotic (another back-and-forth motion, between disorder and order). The methods for tapping the resources of your mind, so fruitful for writing, will be suggested to you in the chapters about **automatic writing** and about your **log** or **journal,** techniques for "pre-writing" and rough-drafting. But for now, let's practice distinguishing the specific from the general.

The fables, myths, proverbs, fairy tales, and parables that we find in the collected folk wisdom of human civilisation are good models for a general truth supported by particulars, or by detailed examples leading up to a general truth, if you wish. The **moral** (a general truth) of "Thirsty Bird" was given to the reader at the end. Do you think it was well worded? Is it true? Could you express it better? If so, do it here, after you've discussed it in class or within your group:

Could you tell a similar story about a person from your own experience, demonstrating this moral? Consider this a possible writing topic for a brief essay.

Now, let's turn to the sour-pussed fox: no moral was attached to that fable. After discussing it with your peers and teacher, give your own best version here:

Do you know the one about the quick rabbit and the slow turtle (hare and tortoise)? Who won the race, and why? Could you write this story out with some details? But first, what's the moral--in your own words:

Finally, after you've read and discussed in your group the zoo story by Carrie Bolden (adjacent Sample Student Voice), see if you can improve on her moral. Write it in your own words:

So far, these exercises have shown movement from the particulars to the general. Though **meaning** in real thought does not emerge in that straightforward manner, freezing motion for purposes of analysis can be helpful. But let's isolate the opposite motion now. I want you to examine some of the following **proverbs,** sayings, maxims, and words of wisdom or fun and then imagine and discuss with your mates how you might develop them with the help of little stories or specific examples:

- Don't count the eggs until they're laid.
- Birds of a feather flock together.
- What doesn't come from the heart will not reach the heart.
- Burned child fears the hot stove.
- If a computer can go wrong, it will. [Modified Murphy's Law]
- Like mother, like daughter.
- Don't get it right. Get it written. **Then** get it right.
- Stolen fruit is the sweetest.
- The proof of the pudding is in the eating.
- Even a fast printer takes longer than you think.
- Money talks.
- Poverty is a wonderful way to start life if you become rich.
- No matter which line you stand in, it's the slowest.
- The Turtle makes progress only when she sticks her neck out.
- Don't quit five minutes before the miracle.

Add a few of your own and of your group mates':

One possible writing assignment (in a moment I'll give you several to choose from) would be to select a proverb or saying as your **main point** (thesis or general idea) and write a story or essay around it, supporting it with particulars.

As you examine the development in your rough draft, you may find that your proverb (your main point) is modified, amended, adjusted--in effect--**changed,** either just a little or quite a bit. You may begin with one generalization and end with a different one. Like some writers, you may choose to share with your readers parts of your journey of **discovery,** for that is the beautiful part of the act of writing: it's an act of self-discovery. You learn new meanings from within yourself.

In other words, the process of going from chaos to clarity is not all slush; much of it is rewarding not only for you to look at but may be so for your readers, too. Great authors like Darwin and Freud, for example, wrote their books by sharing step-by-step their discoveries with their readers.

WRITING ASSIGNMENT:

Unless your teacher tells you otherwise, choose **one** of these topics. In all cases, imagine that your readers are your class mates and your purpose in writing this essay is to make it reflect something of your own interests and character. It's part of your introducing yourself to the class. (Suggested length: at least one page of computer print-out.)

1. Pick any one **fable** that you've ever heard or read, one that you enjoy and whose moral you find significant. Then rewrite it in your own best words. Supply as many rich details as you can.

2. Do the same with a **fairy tale** of your choice. (Suggestions: "Little Red Riding Hood," "Snow White," "Cinderella," "Jack and the Beanstalk," "The Sleeping Beauty."

3. Do the same with a fable never known to mankind before, one that is purely **your** invention (like Carrie's zoo fable).

4. Write your favorite animal story. Be sure to have a clear main point.

5. Pick one of the proverbs or wise words in the list in this chapter (including the ones you added yourself). Then write a story or essay around it, supporting the gist of the generalization. You could also, if you prefer, disprove or disagree with your chosen saying. But do not sit on the fence. Take a stand.

Unless you already have experience on the computer, you'll probably want to write a rough draft on your yellow pad and then later enter it on the screen for revision, editing, and printing.

Good luck on your adventure in discovery of meaning!

Intermission--Sample Student Voice:

Carrie Bolden / English 090-400 / Oct. 1, 1984

A Day at the Zoo with the Mouse and the Elephant

It is said that elephants are scared of mice. So the mouse in this fable uses the frightened elephant to his own advantage. Listen to this.

One day at the zoo the elephant was in his cage eating peanuts and drinking water while watching television--Dumbo, of course. But the mouse was hungry and didn't have anything to eat, and no TV, so he remembered the saying about the mouse and the elephant. The mouse went into the elephant's cage, tapped on his toe and said, "Excuse me, may I . . . ." and before he could get his sentence completed, the elephant had run in fear so fast he broke through the cage.

The elephant ran to tell the zoo keeper, "There's a mouse in my cage, and I want him out!"

The zoo keeper said, "I'll see what I can do, but until then I'm going to give you a tranquilizer to try to calm you down." Later the zoo keeper went to the elephant's cage and, sure enough, there was a mouse in the cage.

The mouse said to the zoo keeper, "You should've seen that big sucker run. I was laughing hysterically from watching him scream and run so fast because he saw li'l ole me till I didn't know what to do. Since he left so abruptly, I didn't get a chance to ask him if I could have some peanuts and water. So I helped myself and since the TV was on I turned to Channel Five to watch Mighty Mouse. Now I'm leaving because I am satisfied, but please tell the elephant that I'll be back tomorrow."

"All right, but next time send some type of advance notice to warn him so he won't tear up another expensive cage. That's going to come out of my pay check, you see. OK?" The mouse agreed.

The zoo keeper prepared another cage and brought the elephant back. On the way to the cage the zoo keeper said to the elephant, "As big as you are, how could you let a little mouse like that scare you? You're a disgrace!"

The elephant said, "If the mouse hadn't caught me off guard, I would have crushed the little devil with my toe. I'm not scared of nothing, especially no mouse, get that straight now! And anyway, my mother always told me not to bully anybody smaller than me."

The zoo keeper said, "That's funny, because you sure tried to step on my toes and stab me with your tusk the other day when I was late with your nuts and with your TV Guide. And I'm smaller than you."

MORAL: People will make any excuse to save face.

9 / YOUR GROUP: WRITING ABOUT PEOPLE [narration]

In this and the following chapter (on feedback) we'll work on techniques and types of writing that will help you benefit from the group method. As you get to know the people in your group, you also begin to practice writing and telling **stories about people**--perhaps including yourself.

To function well as a group, you need to get acquainted, of course. For this I recommend a "values auction" game, an entertaining group exercise that makes you take a close look at your own values--and those of your group mates. There's nothing childish, however, about the play-acting in this auction; whether you're a teen-ager or a mature adult, I'm sure that later you'll look back on it as a serious, valuable learning event.

VALUES AUCTION

A self-inventory of our values can be helpful to all of us at any time but is especially important to a writer upon starting a new project. The simulated auction game in this session is both amusing and clarifying: you might shock yourself when you discover that some of your self-perceptions have been unrealistic.

Everybody in the group starts out with $50,000 to be used for bidding on the VALUES to be sold at the auction that'll begin in a moment, led by your chair as the **auctioneer** (unless someone else is preferred). Your $50,000 is of course useless outside this simulation, so you might as well plan to bid it all. The idea is to bargain for what you really value the most.

Before the auction begins, take a minute or so and go through the list of VALUES on the next page. In the space to the left you might want to put down some remarks to guide you during the heat of the auction (such as "if can get it cheap," "bid plenty", or whatever.) In the space to the right you put down the results: who the highest bidder was and how much you yourself bid on that particular value. You have to keep honest track of your money, too, by making notations. For example, if you are the highest bidder with $25,000 on the value of FAMILY, then you have only $25,000 left and that's the highest you can bid again.

Let's get started. Your auctioneer might start out something like this: "Ladies and gentlemen, the first item for sale on this beautiful day is perfect HEALTH. Do I hear an opening bid? Who will bid $5,000 . . . . ?" And so it goes. If you're not clear on how to do an auction, ask your teacher for assistance.

You'll find the chart of VALUES on the next page.

HEALTH

:
:
:
:

SPRITUAL GROWTH or RELIGIOUS VALUES

:
:
:
:

MONEY or WEALTH

:
:
:
:

FAMILY

:
:
:
:

EQUALITY, JUSTICE, & FREEDOM

:
:
:
:

PROFESSIONAL EXCELLENCE

:
:
:
:

SEX

:
:
:

FRIENDSHIP

:
:
:
:

LOVE

:
:
:

To get the greatest value out of this auction, talk about the following questions with your group mates after you're through with the simulation. Jot down some notes in the space provided before you begin the discussion. Later your teacher may ask each group chair to report to the whole class for further discsussion.

1. Was the game heated or gentle? Explain:

2. What surprised you the most about your own values?

3. About the values of your group mates?

4. Will you be more aware in the future about your own values when you write essays, reports, or term papers? How?

5. Will it help readers if writers always clearly state their own position on any value relevant to their subject? Why?

6. From this auction game and your discussion afterwards, can you see any potential topic for an essay that might interest you? Make some warm-up notes here toward an essay by you on social justice, romance, or whatever **you** deem important. Then confer with your group mates and your teacher.

Veronica L. Jackson / English 090-103 / November 3, 1988 / #5

THE LOVING GROUP

My group is known as the Part-Time Writers but I feel by December we will be known as the "Improved Writers". We should improve because this class is constantly writing summaries, responses, journal entries, group reports, feedback sheets, automatic writing assignments, and essays. This essay is based on our group and the type of love that interests us the most. There are six writers in our group and I'm going to begin with the male population of the group.

The type of love that interests Tariq is his love for knowledge. Tariq feels that knowledge is very important for his survival. One of his reasons for loving knowledge is that it helps him communicate and answer questions on certain subjects. He has studied a vast number of reference books on the subject of computers. This will insure him confidence when relating to others on this subject. Not only is the love for knowledge important to him, but the love he has accrued for the computers.

The other male writer in our group is Albert. The type of love that is important to Albert is the love for life. This love is very general in one respect but he gives specific feelings. He believes through love for life all other types of loves fall into place, such as family love, love for your spouse, the love for the sunrise and sunset. Everything that inspires him springs out of his love for life.

Albert also feels it is impossible to exist without this love because it acts as the foundation of the world. I thought this type of love was unique and this little poem sums it up. "Love sustains every thing; the sun, the moon, the stars the plants, the universe. It's an invisible force that holds everything together,for love is life."

Now that you have heard from the male point of view lets here what the ladies have to say about love.

The type of love that Sivetta feels strongly about is the love that she has for her son and family. Sivetta feels, because of her family and the love they share,that she has been taught how to love and the importance of it. I can relate to this statement, because not all people have learned the importance of love. Since she has become a mother, the love for her family has increased more than she would ever have expected. Her son has made her family more important to each other. She also feels that the pride that her family has in her gives her inspiration to strive for the goals she has set before herself. Even though there are times when her family and she have misunderstandings and disagreements, the pride, love, and ability to communicate conquer those problems.

The love that Sivetta and her son share is her source of survival. The love that she and her family share is her pillar of strength. With out the two she would not be where she is today.

Maritsa, like Sivetta, also has strong feelings for family love. She chose to compare the love for her family with the love for her boyfriend. In order for Maritsa to succeed in life she feels she must have family love. The love and encouragement she receives from her family has helped her learn who she is. She realizes that even if she does not succeed in achieving her goals, her family will always be there for comfort.

The love Maritsa has for her boyfriend is also extremely important. Since he is not part of her immediate family, he has to get to know her and accept her for who she is. Unlike her family her boyfriend does not have to be there when disappointments come, yet he is.

At times this love is stronger than the love for her family. But on the whole the love for her family and the love for her mate go hand and hand. They often get in the way of each other, but the bottom line is

[continued on the next page]

the family love has shown her how to express her emotions. The love she shares with her boyfriend is a result of that family love.

More feelings about love through the eyes of the fifth writer come from Alicia. She has a carefree and open feeling toward her love for life. Her purpose for living is to love everything and everyone. She feels God has given her the gift to be in this world, so she expresses that gift through animals, flowers, trees, people, and everything around her. Along with the love for everything her family is very important. The love Alicia's mother has for her could not be comprehended by her until she had her own child. Now she realizes that the bond between a mother and her child surpasses all explanations.

Alicia feels it would be impossible to live without love. She could just picture herself lonely in this huge world without anyone to give love to or receive it from. Alicia quotes "I would rather be dead."

Finally, we come to the author of this essay, me. The most important type of love to me is the love that God has given me to give to my family. This love is important to me because if God didn't give it to me, I would find it hard to give constant and consistent love to my family. It's not always and easy job being a wife, a mother, going to school, going to work, and trying to maintain a household. After a busy day of all these functions I really don't have the energy to give my family the kind of love they need and deserve. But God always gives me a little extra strength. I'm not saying I don't get frustrated, disgusted, and angry, I think we all do at times, but when I do his love prevails.

This love would effect my pride in a different way than most people would think. I'm very proud of my family. I take pride in knowing that if we didn't have anything to eat or any place to sleep, we would still have each other and the love would be strong. Not that the kids would fully understand and be happy with a situation like this but I feel we would surely manage.

I couldn't live without this love if I tried. I know my existence would not be the same. The reason I say this is because I am very impatient and have a big mouth which gets me into trouble. If this love didn't exist there's no telling what type of actions I would have taken against my family. For example, one time I was sick with an infection and my husband was out of town. My oldest daughter, Verkeya caught chicken pox. She was uncomfortable and grouchy. My son was only about four moths old and still waking up in the middle of the night. When the next morning arrived I was extremely tired and in pain. By now everyone wanted my attention. My son, Keynon, wanted to be changed and fed, and my younger daughter, Valen, was crying and wanted to be fed and Verkeya wanted relief from the itching. Guess what I wanted? To scream, cry, and just leave them there by themselves. But I didn't. That voice said to me, "Ronnie get some help." So I did. I called my mother and she came to my rescue. If I didn't have the love of God in my life, I might have done something crazy by now.

That's all the love I have to share through the eyes of the Part-Time Writers. I hope you have enjoyed hearing about our loving group.

* * *

From reading this essay, did you get any ideas for your own essay about your group? Jot them down here!

INTERVIEWING

By now you have done a fair amount of talking to your group mates and enjoyed it, too, I hope. And that's all to the good. Several aims are served by communal discussion in your group. Yet to get the most out of the group method, we need to **organize** communication so that it will lead to collaborative learning.

Since writing could be said to be like speech that has been translated and preserved on screen or page, using speech or talk in connection with writing makes sense. (I talk to myself aloud while I write, or to anyone in the room.) But **idle talk** in your group--except for the first session or two, perhaps--gets in the way of sound writing practice. You should always have an **agenda** (the chair's job), even if it's only writing on your own. Your teacher will help you plan your agenda, of course, but you should know that writers benefit enormously from being able to take initiative (every new sentence is an initiative!), so your exercising discipline or even leadership in your group will ultimately carry over to your writing and improve it.

The agenda for today is conducting among yourselves, leading up to an essay that springs directly out of this activity. Here are some of the benefits that will be yours as a result of your interview experience:

1. Your group mates would make a good **audience** in your future writing in this class since you've come to know them quite well through the interviews.

2. You'll gain technique in **information gathering.** (For some term papers you'll have to interview knowledgable sources.)

3. When **you** are interviewed by your mates, you'll gain **self-knowledge,** which is most useful for writers.

4. Finally, if you choose (or are assigned) to write an essay about your group, you'll have the feelings that you need--the source material--at your finger tips.

The way to find out about people is, of course, to ask them questions, and then **more** questions. To avoid a coffee klatch, however, we'll now organize the question/answer session. You need to agree on your **purpose.** I recommend that you aim to write a portrait or biographical sketch either of **one member** or of the **whole group.** You don't need to make your choice until after the interviews are over. Or your teacher may decide for you. (Later I will suggest other topic choices related to this session, too.)

When you've agreed on a purpose, make a list of questions. Do not follow your skeleton list of probes too slavishly, however; your list should be a guide, not a harness. As soon as a fruitful area opens up during an interview, be flexible and follow that new line of questioning. Keep asking **more and more specific** questions. If somebody responds like this, "I had a happy childhood," what do you say next?

Before you make up the list of questions, let me warn you against getting lost in the "personnel record file." Some students accidentally fall into that trap and are never heard from again. Their essays resulting from the interviews go something like this:

> "My friend's complete name is .... She was born on .... Her home town is .... Her parents' names are .... Her social security number is .... Her major in school...."

And so on. This kind of material is of interest to no one--except a personnel manager. **Who is your audience?**

To avoid this dangerous snare, all you need to do in your group is to select a unifying, specific **theme.** Naturally you'd want one that is as interesting as possible to most or all of you. The reason it's essential for you to limit your material is that you're not going to write a book. And that would be necessary to narrate the whole life story of one person, not to mention of all your group members. So you're going to **focus** on a specific topic related to all of you.

Once a group wrote about the astrological signs of each member. It turned out to be a well organized piece. I don't happen to be interested in that topic, but that was irrelevant because I was not the intended audience whereas all the group members (and other groups, too) were intrigued by these mystical star-readings. Another group designed a **marriage contract** based on all the opinions expressed during the interviews about how a marriage ought to be structured. Yet another group focused on some of the **values** dealt with in the auction.

When you choose a theme or focus, it could be an age-old human concern, such as "who's the boss in the family?" or it could be a more immediately provoked issue....

> Today is January 28, 1986, and just now, as I write these words in this text, I learn of the spaceship explosion, of the tragic death of American astronauts and their passenger. I'm not necessarily recommending this as a topic for your group essay (though it might be right for some of you). I'm giving it as an **example** of a powerful current subject matter for your group to talk and write about. Look around you. Do you see any issues in your school, your community, or in the world--anything being dramatized right now--that could be the focal point in your interviewing your mates?
>
> Let me stay for a moment longer with the sad example I just introduced to give you some ideas for how group members could deal with a focal issue. In the wake of the tragedy, many

people both famous and "in the street" were interviewed by the media and asked for their emotional or intellectual reactions. The Librarian of Congress said he wanted to cry. A test pilot and best-selling author said he was used to catastrophe and that the space program would go on. An unknown woman said it was madness--hubris--for humans to travel into space. A famous evangelist thought this event, like Pearl Harbor, would bring people closer together. A Harvard child psychologist worried about the effects on all the children watching the terror unfolding on their TV screens. An unknown man talked of the heroism necessary in conquering space. A cardinal offered immediate prayers for the families, also saying that death is part of the human condition. A novelist responded to the interviewer with confusion. And the next day the **New York Times** headlined its lead editorial: "The Death of a School Teacher."

In the same way--whatever the **theme**--your job is to record the whole range of varied opinions and feelings expressed by your group mates during the interview session. Then in your own essay, you may want to generalize about all those specifics so that in the end you can **characterize** your group as a whole.

Here follow some preliminary questions that might inspire you to think of a focus for your own group's interviewing sessions (if it goes well, you'll need more than one session):

Where do your parents and grandparents come from?

Describe the place where you grew up?

Any funny stories from grade school?

What's your career goal?

What's the worst thing that ever happened to you?

And the most wonderful?

If you were dictator of the world, what would you do first?

How many children do you want and why?

What are you most proud of?

Most ashamed of?

What kind of person do you fall in love with?

Now on the next page, make up a suggested list of basic questions for your group:

1.

2.

3.

4.

5.

6.

7.

8.

9.

Let's start the interviewing. The best procedure by far is for all of you to interview one member at a time in the following manner. (Remember to be kind and gentle--this is **no** interrogation!)

You can **begin** with a volunteer or you can follow the roster you use to decide who is the group recorder for the day. The person sitting next to the first interviewee (clockwise) asks the first few questions, adding progressively more specific ones, while the rest of you listen and **TAKE NOTES** on your yellow pads, silently deciding what you want to ask when your turn comes, in order to clarify and deepen the information. Since your ultimate aim here is to produce an essay, the note-taking that I just referred to both in CAPS and **bold face** obviously is paramount. These notes are the beginning of a first draft.

Then you go around the semi-circle of your group until each one has queried the first person to be interviewed.

Next you repeat this process with all the other members.

ooOOOoo

After the interview, do not forget to make some autobiographical notes: how did you feel **when they asked you all those questions? How were you able to respond? Did you surprise yourself with any of your responses? Was it ever difficult to be honest? Could you learn anything about how to ask questions from anybody in your group? Did you forget anything?**

ON THE COMPUTER:

Some students like to pair off and do some of the interviewing with keyboards and screens in the lab. (Remember, the method was described in Chapter 6 / AUTOMATIC WRITING.) The disadvantage is that the whole group cannot be together conveniently; you can do it only in a pair. But the great advantage in doing the interviewing on the computer is that your **notes** can be SAVED to disk and called back later and incorporated in your essay for later revising. So it would save time. Your group and your teacher will work out the logistics for this together.

The system that my students have worked out is this: they do the basic interviewing work with pen and pad in the classroom. Then later they do **supplementary** interviewing in the lab, filling in gaps in their notes. Often as we write our drafts, we realize we need more information from somebody. Don't hesitate to return for more facts.

HOW TO USE THE INTERVIEWS FOR WRITING AN ESSAY:

When you have collected all your notes from interviewing your group mates, I suggest the following topic choices for turning that information into an essay. (If you can come up with an even better one for yourself, confer with your teacher.) Unless you have a better idea, your audience is your group or your class or the student body.

[Normally you'd pick the one topic that appeals to you most for the essay that comes out of working this chapter. But depending on the amount of time available in your course--and on your teacher's instructions--you may write two or even three essays on these topics.]

1. GROUP PORTRAIT: This would be a **portrait** of your **whole group,** properly focused on a theme as discussed earlier in this chapter. (Here are a few more suggestions: "attitudes toward music," "attitudes toward love and marriage," or "attitudes toward the use of computers.") [NOTE: Unless your teacher says otherwise, you don't all have to write on the same topic.]

2. PORTRAIT OF A GROUP MATE: This would be a fairly detailed character sketch of one of the people you interviewed in your group, perhaps the one whose experiences or ideas fascinated you the most. Again, you want to limit the topic in some manner since a complete life story of course would be too long. Consult with your chosen mate.

3. AN AUTOBIOGRAPHICAL SKETCH: This is your way out--if after all this interviewing you still feel that you yourself are the most fascinating character in your group--to write about yourself. To stop from going on forever, however, just as in the other two topics, you need to pursue a theme or specialized focus. You need an angle. What makes you special? What did you learn from what the others in your group asked about you?

Finally a slightly different approach:

4. PROFILE FROM THE PAST: If your experience in these group interviewing session(s) has powerfully brought your mind back to memories of someone important to you, perhaps you ought to write about that someone. This is a **memoir,** which draws a portrait by narrating your experiences with someone you used to know.

Here are some suggestions:

° the best friend
° the worst boss
° the most fascinating stranger
° the most favorite relative
° the most incredible crook
° the most distressing person
° the most lovely child

Whatever choice you make, be sure to consult with your group mates and your teacher before you complete the writing. GET FEEDBACK!

Now grab your pen and pad or fire up that computer and write about the most exciting topic in the world--**people.**

Intermission--Sample Student Voice:

Andrea Davis / English 090-513 / October 15, 1985

Life, Death, and Friendship

The world is a harsh one. Just getting through with everyday life can be trying. Every day people find themselves in some sort of confusion or despair. My own ever-present despair is over an ill friend. He is thirty-six and was recently diagnosed as having leukemia, a form of cancer. Because he is in frequent need of blood transfusions, he is afraid of the risk of AIDS. I wish that somehow I could gather enough courage to let him know that I would gladly give him my blood to sustain his health and life. My blood type is 0 which is the Universal Donor. There is no doubt in my mind as to whether he is worth dying for. He is. My life would be his life. I would be living and sustaining not only my own life but also the life of another human being. I would feel quite useful, for I would have a purpose in life: to give life to a dear friend.

I believe that death is a beautiful thing. It is from the dead that life is born. For example, a dead and decaying squirrel becomes a rich fertilizer for the plants and flowers that will grow and flourish in the very spot where it died. Death is not something to be feared; it is something to be understood. Anything that someone truly believes in is worth dying for. I would feel honored to be able to say that I have a cause for which I´m prepared to die if necessary. Death teaches the meaning of life and not mere existence.

Yet I, for one--being very young--do not yet understand all that death teaches though it has already taught me that I should not take time for granted. I have not taken the time to be with my friend as often as I could have. I did not hug him and tell him that I loved him dearly. I did not act, and now that time is gone forever.

IMPORTANT NOTICE

Ω Ω Ω Ω Ω Ω Ω Ω Ω

For best results in a group-writing course on computers, it's crucial that discussions in your GROUP are at all times about writing in general or about somebody's particular piece of writing. You already know that, of course, but it's a point worth stressing because many times your group may start out discussing one of your mates' essays--yet before you know it, you have all left the essay to its own fate, and you're off in the blue yonder in a wonderful bull session. Sex or baseball would be a good topic for group discussion **ONLY** if it's the subject of an essay that you're analyzing.

You see, you have not been constituted as a general discussion group but as a writer's workshop. Let's say that you have decided to examine an essay written by one group member about corruption in local government. You must then try to remember--especially the chair--that the issue is not corrupt government but how the author wrote about corrupt government. When you're offering feedback on an essay by one of your mates about the chaos in the refrigerator, you're going to dwell not about the mess in your own icebox but on the exact nature of the development and editing of this particular essay which happens to be about chaos in a refrigerator. So, the discussion is about writing.

Always stay very, very close to the text you are examining; otherwise you run the risk of wasting everybody's time and energy. The business at hand is writing and getting better at it--nothing else.

10 / FEEDBACK

Above all, writers crave **response**--or as it is called today in our computer age--feedback. This desire is not just a self-centered whim but a necessity for effective revision of the writing and sound growth of the writer. I believe all but the most experienced professional writers benefit greatly from reader comment.

That's one reason you sit in groups.

Whenever you're wondering how **clear** or convincing or interesting your draft appears to other people, ask to read it to your group. You normally receive much more helpful response after **reading** your writing aloud to your group than after your mates have simply read it silently to themselves.

What if you don't like criticism? What if you are afraid to respond to the writing of others because you might hurt their feelings?

These are common and legitimate concerns; if you're not a literary critic or an English teacher, it's not easy to know how to respond to people's writings. Even professionals sometimes don't feel sure of themselves.

For that reason this text has a couple of specially designed "feedback sheets," which will help you make your response-giving very easy. You can see them on the adjacent pages. (Your teacher may have similar ones for you, too.) The point of using these forms is of course that it **formalizes** the response by putting it in **writing.** Remember, this is a writing class.

The first one you should get acquainted with is called **Basic Feedback Sheet.** As you can see, it's divided into three main parts: A, B, and C. Let me explain the funciton of each of them:

APPRECIATION:

Always begin by offering some **praise** (even if faint). This reinforces the positive tendencies in the writer's work (you catch more flies with honey than vinegar). Above all, you're sitting in **support** groups, so you want to support each other's efforts.

"But what if I just don't like anything at all about it?"

If you truly find nothing to like about a piece of writing, then leave the "A" space blank. If you're having an oral discussion in your group, and the writer asks you pointblank why you have nothing to say (not likely), then simply offer this: "Frankly, this piece of yours didn't do anything for me."

TO:

(name of author)

BASIC FEEDBACK SHEET

A. APPRECIATION: What I liked most about your writing was...

B. SAYBACK: It seemed to me that your **main point** was this:

C. ASKBACK: But not everything was clear to me.

For instance, could you **explain...**

I'd like it if you added some **details or EXAMPLES** about...

Perhaps you could **shorten or cut** the material about...

Additional remarks:

Name of reader:

But in most cases I'm sure that if you try to look for something to like, you'll find it: honesty, sense of humor, telling detail, surprising fact, or realistic description--if not the total work.

SAYBACK:

This is where you give back or "say back" in a nutshell the **meaning** of the essay as a whole. This is a profoundly helpful message to writers because from it they learn whether their writing was CLEAR like spring water, or not. If you don't have **clarity,** you have misfired and must try again. [See if you've heard this expression:"If at first you don't succeed--then you're doing about average."]

Usually the **sayback** that you receive checks out well with your intention. Then you can be grateful to learn that your main point was clear. But what if one or more of your mates are confused by the development of your ideas? Beware the temptation to dismiss their views. ("Other people just don't understand me....") Beware the temptation to argue, overexplain, and argue some more. Just accept the **sayback,** possibly ask for some clarifying remarks, and say, "Thank you." Accept the fact that one or more poeple didn't fully understand your **writing.** It's no reflection on your quality as a human being, only on the clarity of your writing.

Now, when **you** write down your sayback, be sure to leave out any judgmental evaluations (likes and dislikes). Just return--"say **back**"--the message as you received it in the essay, as you saw it, and as you heard it in a summarizing sentence or two.

ASKBACK:

The last section of the basic feedback sheet is divided in four parts. This is where you **ask for more.** Or less.

As you know, one of the chief virtues in effective **development** is an abundance of specific **details** and exciting **examples.** Unless you've just heard or read a nearly perfect piece of writing, you're normally bristling with a need for more examples and details to help you illuminate the main ideas. **Ask for what you want.** Ask for explanations, ask for particulars. Ask for color, facts, figures, and feelings. Your questions are immensely helpful to the writer as a preparation for revision: what should be added?

Or perhaps something should be subtracted. You might ask, "Do you really need three examples of the same idea on page one? It seems repetitious to me." This is also askback.

There are numerous ways of offering feedback, of course, and you don't need to limit yourselves to appreciation, sayback, and askback. Your teacher and your group may come up with further suggestions for how to respond to a writer's work. But when you start out, you'll find our form useful as a model, and in the beginning you may want to use copies of it.

TO: Tyrone
(name of author)

BASIC FEEDBACK SHEET

A. APPRECIATION: What I liked most about your writing was...

I was shocked Ty. That's good because I like to read stuff that shakes me up. I enjoyed the ~~thrills~~ suspense you set up at the beginning.

Your thinking is clear. No doubt about where you stand. And your language is simple but powerful.

B. SAYBACK: It seemed to me that your **main point** was this:

Weren't you saying that you'd die for your family anytime but you'd risk your life for nothing else?

C. ASKBACK: But not everything was clear to me.

For instance, could you **explain...**

Are you often in a situation where guns are used like that? Do you have a gun yourself? I'd like to read more about the background of using guns in your family and ~~community~~ neighborhood. (if that's not too personal).

I'd like it if you added some **details or EXAMPLES** about...

You tell the story quickly in few words—the whole essay is like a gun shot. I like the simplicity. But, but... All the time I wonder about your feelings. Fears? Could you add about how you ~~fear~~ feel? I'm hungry for more details of your experience.

Perhaps you could **shorten or cut** the material about...

No. I want more, not less.

Additional remarks:

What do you mean with "No way, Jose"?

Name of reader: Leslie

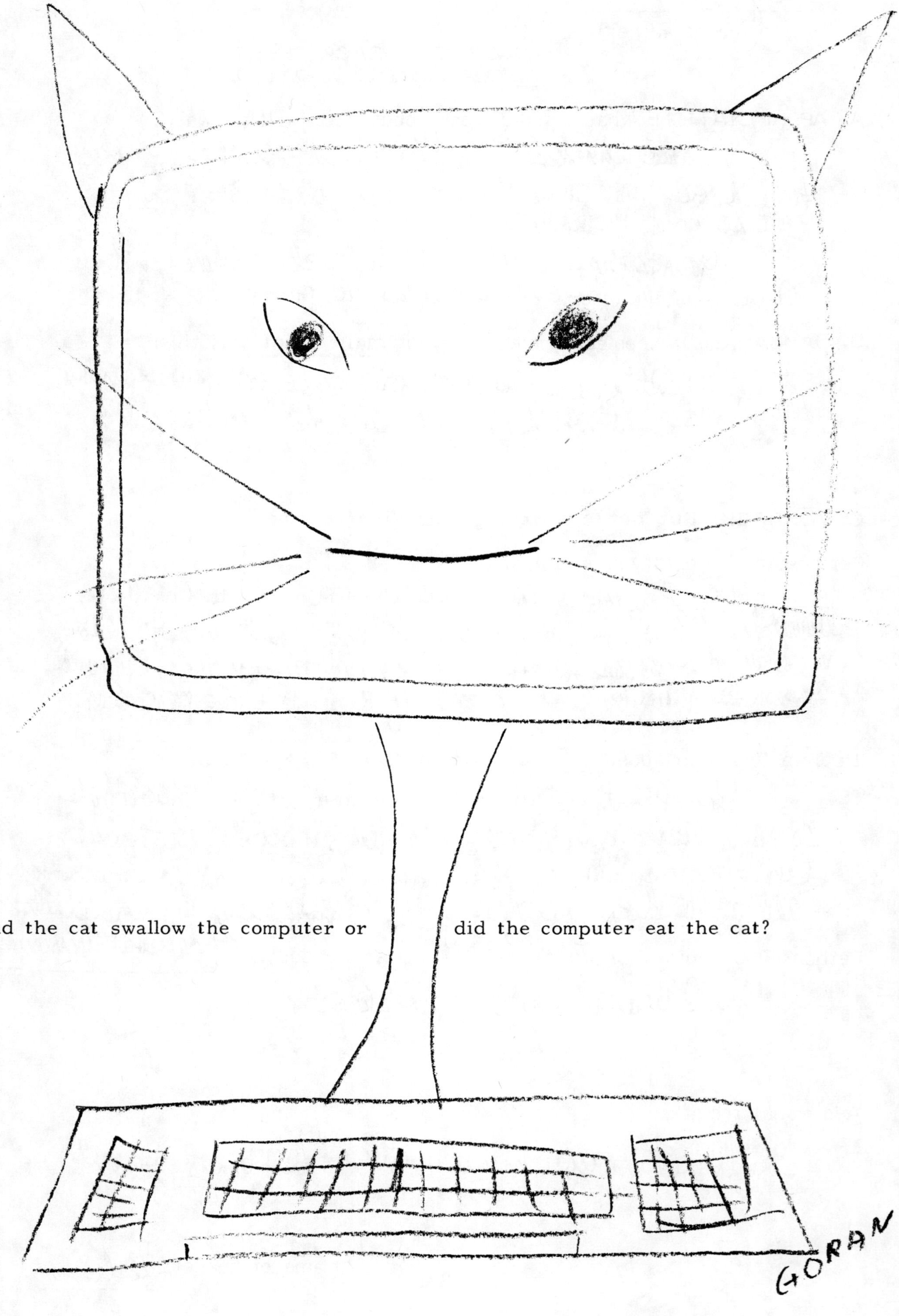

Did the cat swallow the computer or did the computer eat the cat?

NOTA BENE: this feedback sheet is concerned with **development** only. For helping each other with editing for grammatical correctness, see the chapter on Editing Aloud.

GROUP PROCEDURE FOR FEEDBACK:

Discuss this list of four levels of how to do it, from the least good to the best:

° **Inferior:** You give your draft to a mate asking, "Tell me what you think." After a while you get it back with this grunt, "It's OK, I guess."

° **Ordinary:** You give your draft--**with a blank feedback sheet clipped to it**--to one of the group members whose opinions you've learned to respect. You request that the form be carefully filled out. (Here's a variation on this plain-vanilla type of feedback: you give your mate your essay to take home and then return the filled-out sheet the next class.)

° **Superior:** You tell the chair that you would like some time from the group, say about 20 minutes. When this has been agreed to, you **either** hand out copies of your essay draft to each mate (ask your teacher for help with duplicating), **or** you simply read your essay aloud to the group. Then they all fill out their feedback forms and give them to you.

° **De Luxe:** This is the same as **Superior** except there's more to it: you **both** hand out copies **and** read your writing aloud, preferably **twice** (the first time they just listen, the second they take notes for their feedback sheets). Then at the end you have some oral discussion as well.

If time were eternal, we'd only have De Luxe jobs. But when the group can't spare the time for the best, don't knock the other levels (except Inferior). Receiving an Ordinary filled-out sheet from one person can be very useful at times, especially the take-home variation. And now it's time for the computer.

FEEDBACK ON THE COMPUTER:

In fact, the Ordinary level of feedback is often quite suitable for the computer lab where you can turn to a group mate sitting at the machine next to yours and ask her to read your draft either on the **screen** or as--**printed**--and then to fill out the feedback form. Even better would be if your chosen mate could do the feedback on his computer and give you a print-out. Your teacher will have some ideas on this, but you'd also want to experiment in your group until you find a feedback-system that pleases you all.

The Feedback Machine: Here's a method that has been successful in my classes, though It works only when we have a computer to spare, one that's not assigned to any student. We then set it aside for feedback. It works like this:

> A student approaches me and asks for comments from the class on an essay. We then put that student's disk in the drive of the **feedback machine** and call up the essay in question to the screen. That piece of writing is now public and everybody is invited to sit down by that computer, scroll through it and read it, and finally to write a brief (or long) comment at the end of the document. This can follow the pattern of the feedback sheet or it can be "free style"--ask your teacher.
>
> WARNING: Be sure that you know how to **scroll** correctly, without accidentally erasing anything.
>
> There's more. The most exciting part is often when some "critics" return after their initial feedback, read what **other "feedbackers"** have written, and add their further comments. What you get, in other words, is feedback on feedback--a thorough class discussion. (Obviously, if you have a class of 25, there won't be time for everybody, so I run this activity on a volunteer basis.) Finally, the lucky author of the essay on the screen of the feedback machine gets an exciting print-out of many people's honest comments. And sometimes when the united feedback is excellent, I make copies for everybody.
>
> If your group has a spare computer, this can be done as a group-rather than class-activity, too; you simply limit participation at the feedback machine to your group mates.

THE FEEDBACK SHEEET ON FOCUS

When your group is ready for some slightly more technical feedback on development specifically directed to paragraph organization, you can turn to another sheet (see sample on adjacent page). It's not difficult to use, just more specialized and detailed. It requests far more **specific** responses than the Basic Feedback Sheet. Use it the same way as described for the earlier form. Just as your readers want your writing to swim in specifics, so you want feedback that is specific to help you revise effectively.

And you know what else--when your mates are too busy or otherwise unavailable to do this form for you, you can fruitfully fill it out **yourself.** In other words, you'll give yourself feedback. Try it.

For ultimately you'll have to become your own critic.

TO:

(name of author)

FEEDBACK SHEET ON FOCUS

I. Read the first paragraph of your mate's essay. STOP! In the space below, write what you think the author promises the reader, what you think he or she is going to do in the essay. Has the author announced a purpose, a thesis, a chief idea, or a focal point? **What is this MAIN POINT?** (Write it here:)

II. Now read the rest of the essay. Stop, think, and decide if you feel that the essay **as a whole** does what the author promised in the beginning. (See #I above.) Does the rest of the essay develop, support, explain, argue, or illustrate the **main point** that the author announced in the first paragraph?

YES, IT DOES.

[Circle the one that fits best.]

NO, IT DOES NOT.

III. Next (after you've made sure that the paragraphs in your friend's essay are **numbered**), show below **specifically how each paragraph does or does not support the focus** or MAIN POINT announced in the opening:

¶ 1

¶ 2

¶ 3

¶ 4

¶ 5

IV. Does the end of the essay tie in with the opening and **restate** the MAIN POINT? If yes--mark the sentence that does the job or copy it here:

Name of reader

Intermission--Sample Student Voices:

[1]Tanja Pringle / English 090-521 / November 19, 1985

The State of My Room

My bedroom is a mess. In the chair by the door is a stack of sweaters and pants that need to be dry-cleaned; there are about three pairs of pants and maybe six sweaters. A coat that I wore two days ago is also thrown over the chair.

On top of my stereo is a small brown paper bag that seems to be empty. The cassettes that I listen to are scattered about, some out of their cases, some not. On top of my dresser are a black head scarf that I wear to bed, a few hair rollers, a letter that I need to mail, and an article that I had cut out of the paper to use for my computer-writing class. A piece of beige paper is sticking out of the bottom drawer on the right.

On my bed at the far end lie a pair of jeans and a brown sweater that I wore today. Hanging on the left bed post is a green shirt that needs to be taken to the cleaner. Wrapped around the center of the bed post are three belts, a red, a blue, and a brown one. Under my television stand are various magazines out of place, as well as a glass, two bottles of fingernail polish, and my eyeglasses.

On the floor near the chair is a stack of about a week's newspapers that needs to go in the trash. On the floor are a plastic bag I used the other day to carry my shoes to the shoe repair shop and a pair of boots turned over, showing chewing gum attached to the heel.

It's clear that my room is a mess. I promise myself that tomorrow I will clean it up. But not tonight. Right now it is time to go to sleep.

---

[2] Margaret Cox / English 090-521 / November 14, 1985

The Chaos in my Closet

My clothes closet is like an armed camp. Sometimes I am afraid to open the door because I don't know what will attack me. Since I have so much stored in it, it becomes dangerous terrority.

On the top shelf are two boxes of dress patterns, several designer hats stored in hat boxes and small items such as stationery boxes and dangerous electric curling irons. It looks as if every time I want to wear a hat it is at the bottom of the stack.

The floor of my closet is like an ammunition store. That is where I have a large black garbage bag full of dress patterns and fabric. It is my own private store because whenever I need something new to wear, I find a pattern and suitable fabric to make whatever article of clothing I need at that particular time. Also on the floor is a large pamper box where I store my winter or summer clothing, depending on the season. I don't like to pile the floor too high because I need space to hang my clothes on the rack so they will not become wrinkled. I arrange them so the short articles are on one side and the long items on the other. On the door of the closet hangs a shoe bag very loosely. Some compartments are stuffed with two pairs of shoes that sometimes become flying missiles when I open the door. So in the interest of peace I keep it closed most of the time.

11/ WRITING IS SEEING [visual description]

"The writer's task is to make you hear, to make you feel--it is, before all, to make you see."

--Joseph Conrad

To improve your observational power as a writer, hold your pen and your yellow pad in your hands while you **look**, like an artist doing a sketch. And remember, you **are** sketching verbally. Look at your object or scene, then look down on your paper and write, then look up again, and so forth. Some people don't bother to write in sentence form at this stage but simply make lists. Do what feels comfortable for you, but do get something down on your pad.

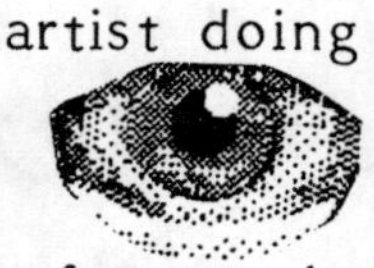

Here are some suggested exercises in visual description:

- Describe your classroom: physical details like colors, shapes, sizes.

- Do the same in your lab: what exactly does your computer look like anyway? Let your description catch the very motion on the screen as you write. This is indeed getting close to the object.

- If there are any hills or tall buildings in your area, ascend with your group and scan the horizon. Describe what you see. What's the overview--can you generlize about it? Then offer support in the form of specific details. Do you have access to binoculars? Zoom in on some interesting details.

- Give a rundown of your "writer's habitat," the area at home where you do most of your writing: the chair, table, parapharnelia, and surroundings (some people write in bed--do you?). Be as **visual** and **detailed** as you can.

Without getting technical and deep into psychology, we can say that human beings **think in pictures** or images. Since writing is thinking put down on the screen or page, the more **visual** your writing is, the more clearly it communicates your thoughts. Next is a powerful exercise for developing your **visual** writing.

Take a close look at the engraving by Gustave Dore on the left page. First get a quick feeling for the total picture. Then start spotting the details. When and where do you think the scene is from? Even though it may be some distance from your own life, do you recognize anything? Have you ever seen anything **similar** closer to home? Pictures are not only "out there"; they can also be inside you.

Take two inventories of this picture. First **list** at least nine specifics that you **see** and then some of the **emotions** stirred by the sensations entering your eyes your eyes perceive in the picture.

DETAILS	EMOTIONS
1.	1.
2.	2.
3.	3.
4.	4.
5.	5.
6.	6.
7.	
8.	
9.	

Wasn't it easier to see than to feel?

Now compare your details and feelings with those of your mates. What did you list that they missed? What did they list that you missed?

What can you learn about yourself (and about others) from comparing the various lists in your group? Did you get any ideas for improving your observational powers? Do you regularly record your visual experiences in your **journal**?

The next step is important. Move away from the details and get a bird's-eye view of the engraving. Hold the book at arm's length. Then below write down a **generalization** of the picture as **total image** to you. Try to sum up in one sentence the visual impression this picture has made on you:

Intermission -- Sample Student Voice:

William Purcell, Jr. / English 090 / Nov. 1989

GUSTAVE DORE'S ENGRAVING: A DRAMATIC DIALOGUE

1st MAN/left): "Hey, buddy, what's the story?"

2nd MAN/left: "I think our neighbor ran out of food. So they put their kids on the streets to rummage up some money or eats."

1st MAN: "Yeah, well, what can they do about it? We're all in the same boat."

2nd MAN: "Anyway they ain't standing around griping about it or waiting for something to come their way. They're going to the Rich Folks town and trying to dance and sing, passing around a hat for contributions. I think I'll go with them."

1st MAN: "Ah, it ain't gonna do no good. Them rich folks always sticking it to the poor, even when they're selling a song and dance."

2nd MAN: "It can't hurt trying."

1st MAN: "So what makes you think they won't throw the poor kids on their 'hinetails?'"

2nd MAN: "Because they're selling something that the rich folks need themselves: feelings, songs, dancing, something they don't know how to do."

1st MAN: "Well, you go with them if you want to. I'll stay here and wait and see what comes around, if you don't mind."

2nd MAN: "Why should I mind? It's your loss."

....

Hours Later . . . .

....

1st MAN: "Ha ha ha, how'd y'all do? Didn't I tell you they'd throw you out?"

2nd MAN: "We all made $75 a piece in cash and got $75 worth of food. To top it off, the mayor wants us to come and sing and dance for his birthday."

1st MAN: "Is that how you want to live, with your hat out scraping and bowing?"

2nd MAN: "Don't deny until you try. Those who seek shall find."

We're going to get deeper into that old picture yet. Take a close look at the various characters, young and old. Then enter the mind of one of them that interests you. If you're a woman you may want to pick a woman; if you're a man you may want to pick a man--but not necessarily.

> Next you grab your yellow pad or fire up your computer and write an **interior monologue**, putting down whatever you think is going through the mind of your chosen character, considering his or her circumstances and facial expression in the picture. Spend about ten minutes on this monologue.

The next part is even more exciting: suddenly two of the characters (one of them is probably the one whose interior monologue you've already recorded) come **alive** and start talking to each other while the rest of the people are listening. What are they talking about? Create a **dramatic dialogue**!

> You can do this in one of two ways. You can be the master of the dialogue and create the lines of both people. Or you can get together with a group mate and do it jointly; in this case, of course, you take one character each. If you've already practiced interviewing on the computer, you'll know how to get together with a mate and write this dialogue on the screen. (This would also prepare you for the writing in Chapter 20, "The Play Is the Thing".)

ooOOOoo

Is the old engraving more alive to you now than when you first saw it? Do you know of any other picture, painting, photograph, sculpture--any piece of art--that you would like to take a closer look at? Select your object, get your yellow pad or your machine ready, and proceed in a manner similar to what you've just practiced here. What you write may become the rough draft for a full-fledged critical essay.

Would you like to become an art critic?

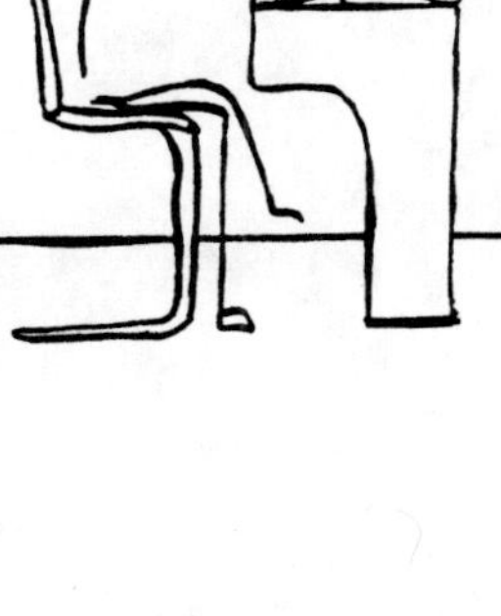

Alex Augustus / English 090-084 / October 6, 1988 / #2

CHAOS IN THE UNDERGROUND

Me, I don't have much chaos in any of my personal stuff, maybe in my private life, but that's about it. I'm a relatively neat person so the only other chaos I can think of at this moment is the chaos on the subway train! Every morning is the same old thing. I walk about five blocks to the train station to get on the "G" train.

The train is rarely on time, so no matter how early I leave my house, I still get to school late. When the train comes, it is so packed, I'm tempted to wait for the next one. I would, but then I'd be later than I am now. So I square my shoulders and ram my way inside the train before the door closes, before I change my mind. On top of that, some people have a body odor that won't quit. It's as if they haven't taken a bath since they discovered water.

At times I really don't know which is worse, the people or the train. The trains are so dirty, to the point that I don't want to touch any part of it. The floor is covered with dirt, paper, food, and stains. I often wonder if the stains are from soda, or coffee that someone accidentally spilled, or is it bodily secretions from a homeless derelict. Also there is graffitti on the seats and across the subway maps. Imagine, a whole fleet of subway cars that are clean. It'll be an Eighth Wonder of the World. Almost every train I see is covered by a layer of dirt.

Once the train I was on wa so crowded that I decided to move to the next car. When I got to the next car, I looked at my hands, and they were so dirty it looked as if I was working on the railroad. I try no to lean on the door if I can help it, but when there are no seats available, and all the handrails are occupied there's not much I can do except to lean on the door. The only problem with that is if you're wearing any sort of light colors, your clothes become a nice shade of gray. Also, there is that chance that the door might fly open unexpectedly while the train is moving. With the AIDS epidemic, and other sorts of diseases going around, I'm skeptical about holding on to the handrails. I know that you can't contract AIDS that way, but still, it's just the idea. The handrails feel slick or greasy sometimes, and I find my hands sliding down the pole. I look at my hands then I look at the pole and I wonder what's on them. Then again I don't want to know what is on them.

Anyway, after I get off the "G" I then walk across the platform to the "A" or "C" train. Again it's the same old thing. I have to fight my way through the sea of people and get on the train. I haven't had so much blocking exercises since I played football in high school. Finally, I get off the train and then rush to school. I catch my breath and take a deep sigh. I say to myself, "I made it." Then the reality hits me--it's only Monday, and I have to go through this for the rest of the week, the rest of the year!

I must get a car.

FROM CHAOS TO ORDER:

The normal process for most writers is a movement from messiness--or chaos--to order and neatness. That's the creative act. Some of us are too hard on ourselves about the mess, forgetting that we have to start somewhere. Without initial disorder, we cannot move forward, gaining progress and striving ultimately for perfection.

Let's **look** at chaos in other parts of our lives for a moment and see what we can do about it. The interesting and useful point I'm going to make now is that the very act of examinaing and **describing** something that is out of order will in fact immediately start the ordering, at least in your mind.

In a brief essay, make some **order** by describing the **chaos** in your

- bag or purse or briefcase
- closet
- bathroom cabinet
- refrigerator
- kitchen sink
- car
- [your choice:]

Two modifications to this exercise:

A) If you are an exceptionally neat person who resents the suggestion that there's chaos anywhere in your life, write your essay as a description of the neatness and order in any of the above areas.

B) A more intriguing and much more advanced assignment would be to describe "THE CHAOS IN MY HEAD," **visualizing** your state of mind at certain times. What do you **see** in your mind's eye? This is guaranteed to bring you some clarity.

THE EYES IN YOUR COMPUTER:

Shut your eyes momentarily and imagine that the face of the person you love most is staring at you from the screen in front of you. Open your eyes and quickly write a detailed description of his or her **eyes.** If you have a hard time, look at the original next time you have an chance and try again. (Your beloved will note your attention and may well return it.)

Intermission--Sample Student Voice:

Mary Ingram / English 090-522 / November 13, 1985

A Visual High

I could write about the chaos in my closet or the emptiness of my refrigerator but anyone reading that essay would get bored after the first paragraph. So instead I'm writing about the varied views from my office window on Broadway in Midtown Manhattan between 44th and 45th Streets.

My office is on the 29th floor of the once-named W.T. Grant Building. I'm not sure what it's called now, probably the new Astor building or maybe years ago this is where the old Astor building used to stand. I get such a great view of the buildings over on the West Side. The agency that I work for was one of the original tenants and during the thirteen years we've been here a lot of new skyscrapers have gone up. I think I must have worked on just about every side of this building. At one time I was facing the East River, and on a clear day we could see New Jersey for miles. And on a very clear day, you could see the cars moving about in the townships far away.

The view that I have now is spectacular. Do you know I can see pedestrians and traffic on 5th Avenue which is three blocks away? It's like looking out on a great big stage where I can see balloons and newspapers flying about on windy days like some graceful ballet dancers. If I could fly, I wonder what it would be like to soar up and around the skyscrapers in the area.

Being up this high I can take a good look at the skyscrapers, and you can't imagine all the many visual impressions available to us up here. Looking down, it's like a tapestry of people, with fabulous patterns and colors. I can also see the Pan American Building and helicopers buzzing the Chrysler Building, dressed in its Art Deco hat. From the office next to mine I can see the Empire State Building. It was exciting when the movie people put up the King Kong gorilla. It was such a sight, the high wind defeating their vain purpose. Of course, on a cloudy day we can't see anything up here.

You should see the window-washers go up and down on their mechanical contraptions to really appreciate your choice of a different profession. Not long ago I had a true scare. I knew that repairs were made to the outside of the building, but not what kind. Then as I was relaxing one lunch hour I heard noise outside my window. Some men where washing the windows and going about their business without a care in the world, with only ropes tied around their waists and standing on a narrow scaffold. Can you imagine being 29 floors up without a safety cage? It gave me such an eerie feeling I had to close my eyes.

Sometimes sitting at my desk I've seen flies crawling across the window and butterflies floating by. Did you ever think insects could come this high? I thought they stayed closer to the ground. Perhaps they, like me, want to enjoy the good view up here of the sun going down. Its reflections in the skyscraper windows are a photographer's paradise. I hope I'll have this view for a long time. Every day I notice something new.

12 / SELF-EVALUATION [progress report]

The purpose of your writing in this session is to give yourself a few moments of reflection on your progress as a writer-in-training. This analysis will benefit yourself, your mates, and your teacher. Don't forget to compare notes with your group mates when you're through.

Follow this process: look at the questions below a few days before you start writing and discuss them within your group. Then make some rough notes with your pen in the spaces below. Finally elaborate and expand on the points you made in these notes on your computer. Your teacher will tell you when the print-out is due to be handed in.

1. JOURNAL: How many times a week do you write in it? Do you enjoy writing in your journal or is it a chore? Give details. Include a few short sample quotations from your journal.

2. **Writing on Computers:** Are you making full use of this book? Are you underlining and writing in it? Which chapters have you found helpful and which ones not?

3. YOUR GROUP: Are the members of your group working together smoothly and effectively? Give detailed examples:

4. THE COMPUTER: Describe your progress. Do any of your computer-writing skills carry over to your writing in longhand? Explain! Have you developed a combination of writing with pen and paper and with the machine? Do you always draft on paper first and then enter the text on the screen or do you also create draft on the computer? What's your best process at this time?

Marie Johnson / English 090-524 / April 13, 1989 / #3

Self-Evaluation

I write in my journal six times a week. Maybe I'll write seven times depending on how I feel and what has happen that day. Sometimes I don't feel like writing but I'll write down just that and keep on writing. To be honest, at first it was a chore but now I enjoy writing in my journal. My biggest problem was starting the journal. I just don't know what to write.

For example, on 2-10-89 I start off like this. "I'm not really sure what to write about. I guess I'll write about my day. It's now nine o'clock and I'm getting myself ready for work. I don't have to be there until twelve o'clock. Today is my late night so I'll have to work from 12 to 8 pm."

Here's another sample of my writing but a different day. 2-12-89 "Life and death. What is life; a day of sunshine or a breath of air. What is death; can it be a day of sorrow or a black cloud. Life is what we receive and death is something that well come. Why do we have life when it well come to death?" I try to write what I feel instead of thinking of what to write. It's not always easy to keep writing but I give it my all.

I use my book all the time, even when I'm at work. It not only helps me in but at work too. At first I wasn't writing in my book. I don't like to mark my books up. By underlining in the book it helps me find things I don't know or was helpful to me. When I write in the book it keeps my thoughts fresh in mind such as in chapters 8, 9, 13, 14 and 15.

My group members are easy to work with. We sometimes fall a little behind, but we seem to pick up right away. Everyone wants to pass this class so we help each other. We help with the computer work and our writing at the desk. We also give evaluations on each other writing. Like with our group essay everyone gave their opinion.

Yes I feel my computer writing skills have carried over in my longhand writing. The computer is easier to move my words or sentences around. With longhand I would have to rewrite the paper. I'd rather do a draft of my wring first. Once I've gone over the draft I can put it on the screen. That's probably my best process right now.

13 / THE WRITING TEST [essay tests, finals, etc.]

So far this course has encouraged you to write freely and leisurely, mostly on topics you've designed yourself. This approach is making you an experienced and probably comfortable writer.

Unfortunately, however, sometimes you may be required to write with a **time limit** and on a topic **not of your own choice,** such as in-class tests with essay topics and other writing assignments like the "one-hour essay". Let me give you some pointers on how to deal with the common difficulty of time-limited writing tests. As always, to learn to do a difficult task, we **practice** doing the difficult part of it.

THE TEN-MINUTE ESSAY:

This is the best trick I know for getting a fast start on a piece of writing. (It has done wonders for hundreds of my students in helping them pass the City University of New York Writing Assessment Test.) The 10-min. essay is not as new as it might sound to you; it is really a modified automatic writing session. You can do it either in the group mode or in the class mode:

1. Your group chair or your teacher, as the case may be, announces the assigned topic(s) and begins the timing--"Everybody **go!"**

2. You begin writing at once, thinking with your pen--or fingers on keyboard--**as if you were doing regular automatic writing.** But instead of letting your thoughts free-float among your idea associations, you do your best to **focus sharply on your feelings about the topic and to seek out examples, examples, and examples** that have a bearing on the main point you think you want to make. I say "think" because you might adjust your main point as you develop your draft. (See the "Suggested Steps" on the adjacent page!) In this manner you avoid the common mistake of "philosophizing" loosely and abstractly in an endless introduction. Instead you get right down to the nuts and bolts without repetitious generalizations. Your readers will love you for that.

3. When "Time is up!" you'll be amazed at the quantity and quality of your writing: many students will write several hundred words--the better part of a rough draft--in ten minutes, much of it salvagable. And you have most of the hour still left for revision and editing.

You will find that if you do a dozen or so 10-min. essays with your group or class mates, you'll become so accustomed to the time pressure, that you will be quite comfortable when the real writing test arrives.

Intermission--Sample Student Voice:

Andrea Davis / English 090-522 / Dec. 16, 1985

Writing Timed Essays

When I started this course, I felt that it would not have much to offer me. This is a "basic" (remedial) writing course, and my writing was above that level. I thought that I might be wasting my time in a class that would teach me the basics that I already knew.

I was wrong. Although I was not in need of overall remediation, I did need to discipline my drafting and also to sharpen my editing skills with techniques like editing aloud and scanning the screen. The reason why editing posed such a problem for me in this course was that our practice writing tests (preparing us for the final) were all *timed*. But prior to this course my experience had always been with *long* essays that usually had not been due the same class session.

This led to a difficulty for me in getting my main idea down on the page. Because I was accustomed to lengthy essays in which my method was to contrast two perspectives in some depth, I now felt constricted in supporting my thesis richly enough in one hour. When I had to write *timed* essays, I at first felt the result was impoverished by a vague main idea, lacking in support. And spending so much of the limited time on development left none for editing at the end.

As I progressed, I learned to shorten my essays and to quickly focus on one main point. I have usually written all my essays by first over-expanding my main point, but now I discovered that in shortening my essays, my writing became more and more concise and straightforward. What helped me the most in reaching that point was all those *ten-minute writing exercises* that we did in class every day.

They have helped me so much that I think I will continue to time myself on my own whenever I have writing assignments. This technique has made me realize that in the past I probably took too much time in writing an essay, using up a whole day or even week to complete it.

I now also feel more comfortable in the drafting stage when I write on paper as opposed to the computer. First of all, I will not always have a computer to do drafts on. Second, although writing on the computer keyboard is a lot easier, I'm truly accustomed to the old paper. Yet for editing, the computer is best because I can spot common errors in verb forms or spelling more quickly and efficiently.

Suggested Steps for a One-Hour Writing Test (for Real or for Practice):

Instead of beginning writing at the top of your yellow pad, start writing about 5-6 lines down the page. Then if you think of a better beginning later, it'll be easy to put it there.

1. **DRAFTING:** When the timing begins, start **writing** at once. Think with your pen. If you know your MAIN POINT right away, great. If not, don't worry. It will come to you in a moment. In the meanwhile, dig right in with **EXAMPLES** that you associate with the topic. Milk each example for as many specific details as you can: WHO, WHAT, WHEN, WHERE, WHY? Don't worry about a conclusion. **TIME:** Spend about 20 min. on this rough draft.

2. **REVISING** (don't confuse this with editing--see below): **TIME:** Take 5 min. or so and re-read your draft, looking for the following:

> a) Is your MAIN POINT clear? If not, improve it. If you find a main-point sentence in the middle of your draft, would it sound better at the beginning? Although a conclusion is not always necessary, simply restating the main point at the end often is a good idea.
>
> b) Do you have enough EXAMPLES and do they support your main point? Should anything be cut? What can you add--some specifics or even a whole new example? **Do not edit** at this stage!

3. Now get up and stretch your muscles, get the white paper, and begin **COPYING.** (Many writers who are winners at the timed writing game, copy while softly reading their writing aloud. This means that the copying really is also editing. You kill two flies with one slap by copying aloud.) Do you want a title? **TIME:** Allow 10-15 minutes for the copying.

4. Finally you **EDIT ALOUD. TIME:** You should have saved 5-10 min. for this most important step.

REVIEW THESE FOUR STEPS IN YOUR OWN WORDS HERE:

1.

2.

3.

4.

Intermission--Sample Student Voice:

Ethel Crawley / English 090-161 / March 14, 1985

Mission Accomplished by The Editors Group

This is a warning to all of you little Stumbling Blocks of Writing, such as Hatred, Sloth, Boredom, and--yes, you too--Period Mistake. You won't be around much longer. For there is a new group in English 090--The Editors--that gets its mission accomplished. So pick up your luggage and get a move on, Hatred for Writing.

With delightful fighters like my group mates, who get their job done, writing is no longer a chore. It's a pleasure. Speed up--you Slowness of Writing--for with techniques such as automatic writing and editing aloud that this group is now mastering, the words are jogging along, the sentences flowing. You there, Boredom, now that Hatred is packing, you too can take your buddy Dullness along on your flight because The Editors are too busy interviewing each other to notice your boring intrusion.

Now it's your turn to get packing, you sneaky little Period Mistake. Don't forget to throw all your demonic Faulty Commas in your bag on the way out. Yes, The Editors are practicing editing aloud together and ready to wipe you all out for good. We won't tolerate your presence much longer; together we can clean out any mistake known to mankind. Our team work will rid us of all obstacles on the way to clear writing. Yes, all of you will soon be known as Stumbling Blocks of the Past.

In our group we work together, giving our best to each other, loving every minute of it. Each member possesses a special touch that enhances our learning experience. We provide information about missed work to any member who has been absent, for example. And we give constructive criticism on the feedback sheets to strengthen our resistance to all of you departing Stumbling Blocks. Yolanda, for instance, has explained how more details were needed in one of her group mates' essays. And Jackie has helpfully demonstrated to a member the technique for placing a jab in the solar plexus of the Period Mistake.

Now, don't get the impression that The Editors are all work and no play. As a matter of fact, our sense of humor is one of the qualities helping us in our struggle against you--we fight with satirical pen in hand and defeat you with **words**. So, dear Stumbling Blocks, you had better beware of Efrain, Yolanda, Jackie, Emerita, Bebra, and Ethel.

We get our mission accomplished.

14 / EDITING ALOUD [comparison & contrast]

Trying to avoid errors in grammar at the same time as we are **writing** is extremely difficult, as you remember from Chapter 5 ["Development vs. Editing"]. It's much easier to write first and later go back and make corrections. But here's the $64,000 question: **how do we detect our own "errors"** (non-standard usage)?

To say that you "edit" your writing by reading it and "checking" it is not enough. **How** is this editing done? You need a detailed, specialized method. Fortunately we have one, a superb technology, though often neglected: **voicing** the reading of your writing to yourself. That's right, you edit ALOUD.

"I guess that's OK at home alone," Joachim objects from his corner group. "But I can't do that in class." Of course you can. The word was **aloud**, not "loudly." As soon as everybody in your group and in the class surrenders to the awareness that the only effective path to self-editing goes through **voicing**, you will find comfort in the gently rising and falling sound in the room, as in a beehive. It's the music of writers at work. Your teacher at first, and your group chair later on, will lead you in regular sessions of editing aloud.

"I had a junior high school teacher once who told us to edit aloud, just as you say," Daniela blurts out, "but it didn't help any."

You said your teacher "told you to"--but did you actually do it, Daniela?

"I guess not."

That explains why it didn't work. The big difference is that in this course you will actually train yourselves to edit aloud in special practice sessions. Eventually it will become second nature and you will be effective self-editors.

"How come this voicing works?" asks Joanne.

Let me explain. In silent editing, you mainly use your visual sense, only one sense. By relying also on your **hearing, you double you mind's power to detect error.** As you keep reading aloud to yourself, from time to time you'll notice that the verbal signals from your eyes and from your ears are in conflict. You note a contrast. You're then faced with a **choice.** Your decision at that moment is an act of creative editing.

Let me give you some detailed examples. As you move down the lines of your text (whether on pad, screen, or print-out) you **read slowly with your eyes and mouth.** That's right, you read with your mouth, too. In this process, you're coordinating three motor systems simultaneously: the **muscles** of [1] your **eyes**; [2] your **voice** apparatus;

These foolhardy people insist on editing in deadly silence.

and [3] your **arms and hands.** It's a three-dimensional job, like piloting an aircraft. As soon as you overcome the initial resistance to the novelty of hearing your mouth doing part of the editing job, you'll feel suddenly rich, and you'll become adept at the process of correcting your own mistakes. It works. Typically you'll discover "conflicts" that you'll have to resolve such as these:

- You'll **see** "they says"--but you'll **hear** "they say."
- You'll **hear** yourself make a definite, long pause at a spot where your eyes **see** no period.

What happens specifically is that as you come upon a word, you feel a need to pause because a second possibility occurs. For example, when you come to "mens," the word also flashes inside your head. You have a **choice.** This means that you will direct the muscles of your arm and hand to make the change that is consistent with Standard Dialect. You will **hear** that although "they says", for example, may be all right when talking among your friends from the same dialect group, the other choice--"they say" **sounds** like what you want in your essay. The same goes for "mens/men" and so forth. Similarly, you'll hear/see missing or superfluous **periods** (run-ons and fragments).

When I first introduced this voicing technique in my classes some years ago, I was amazed at the seemingly magical results. Student-writers who had struggled with subject-verb exercises in drill books for a decade or more, with little improvement in their editing skills, reduced their mistakes by half after only a few weeks of practice in editing aloud.

There's no magic, however. This method works for good reasons. Native speakers of English (as well as most ESL people who have spoken English for several years) have a golden mental resource that they may not even be aware of. Although most people normally speak in their regional or ethnic dialect, they also have solid but **passive knowledge of Standard English Dialect.** And that's what I want you to tap into.

By "passive" I mean that while many people don't normally express themselves in Standard Dialect, they know it when they see it or hear it. Most of the time they can also perceive its absence when it is missing--such as in their own writing. In other words, through training they can activate this passive knowledge and profitably use it in their editing.

"You mean I already know correct English?" Sebastian looks pleased and mystified at the same time. "How long will this take?"

Yes, you have a reservoir of dormant knowledge of Standard English. Learning to activate and exploit this information system during editing times should take you anywhere from a few weeks to a couple of months. Naturally it depends on your discipline and stamina. But once you've acquired this new skill of editing-aloud, you'll cherish it. You'll write with greater self-confidence than ever.

Before I explain the detailed steps of this editing process, let me say a few more words about what it takes--beside discipline and stamina. In contrast with filling in blanks in drill books, training yourself to become a self-editor cannot be a mechanical act (though it may seem so once you've mastered the technique). No, it requires enthusiastic initiative and willingness to **take responsibility for your own writing and editing.** It won't work as well if you do it to please your teacher as if you do it to please yourself as a writer. "I have to correct this for my teacher" does not reflect as effective an attitude as "I want to edit my writing to make it as correct as I can." So what helps most is a genuine pride in your own work, a persistent desire to produce the best.

STEPS IN THE PROCESS OF EDITING ALOUD:

1. Read your writing on your screen or paper **slowly** and **aloud** from the beginning through to the end. Slowly and aloud.

2. As you read, **Listen** carefully to what you hear yourself say while you also **see** what you read, all the time **comparing** the two. When you **hear** yourself say something different (even slightly different) from what you **see,** pause to investigate this contrast, or it may be a word or ending left out. It may be something you don't need.

3. When you see a **choice,** repeat the two possibilities aloud, again and again until you're ready for a decision. Make the change and move on.

4. Do it again. Read through the whole piece again. And again. Three times is the minimum; four or five may be better. You can see how important it is to organize your time so that you will have saved at least ten minutes for **editing aloud** at the end. (For a term paper done outside class, you may spend several longer sessions on the editing work.) To get the most out this process, you had better **focus on a specific problem each time you read through your essay.** Design your own list of what to look for and listen to. Here's an example:

a) Listen to hear if there are any **missing** sounds.

b) Listen for the **pause at the end of each sentence.** This is a sure-fire way to spot and fix run-ons and fragments. Some people have found it helpful to reinforce their listening power by **tapping** their desk gently with their fingers at each period-pause. This way any conflicts become quite jarring. [See also the next chapter, #15.]

c) Check for correct **verb forms** (if that's your problem). [See also Chapter #16]

d) [Put whatever is your personal demon here.]

5. This fine-tooth combing is optional, but I recommend it for every serious self-editor: You go through your writing once more but this time **backwards.** You begin with the last sentence, reading it aloud forward. Then you move back to the previous sentence and so forth until you come to the opening sentence of your essay. You see, with this seemingly crazy procedure you will not be distracted by your own wonderful ideas. Instead you can concentrate on the sounds of each separate word.

That's it. Writers who want to be winners do not skimp on editing practice. (If you think that filling in blanks in a drill book can substitute for the process of editing aloud, think again. Has that activity alone made you a good editor in the past?) Be sure eventually to **design** your own little list as suggested under #4 above.

Ω Ω Ω Ω Ω Ω Ω Ω

Editing in the Computer Lab

The editing-aloud process works especially well when your writing is on the screen or on a print-out because of the easy legibility. In fact, difficulties in reading your own handwriting might have been the cause for quite a few of your errors. Fortunately, it has been the experience of many of my students with just that problem that once they improve their editing with the help of the computer, that improvement carries over to their handwriting work.

Important reminder: Your having edited aloud on your screen several times does not preclude your re-editing the print-out. You now have another chance to catch what you missed earlier.

USING "FIND & REPLACE":

Most writing programs have a function that moves your cursor in a flash to every occurrence of YOUR CHOSEN word, phrase, or even single mark like an apostrophe. Not only that, activating this command--usually called "FIND?" or SEARCH & REPLACE (depending on your program)--will also allow you automatically to **replace** one item with another. This is one of the most amazing tricks of your electronic writing program. It can be used for revision (development), of course, as suggested at the beginning of Chapter 8 about your journal writing. But it also has many uses for helping your **editing** work. Let me give you a few examples.

Imagine that you know you often forget to correct things like, "People was" or "they was." What you do when you're ready to edit your essay on the screen is this:

You tell the computer to FIND every occurrence of "was." You ignore the REPLACE function because you don't know if and how you want to change each "was" until you see it.

Or let's say that you know you tend to leave out the final "**d**" in "supposed to" and "used to" [**wrong:** "We're suppose_ to...; **wrong:** "Back then I use_ to..."]. So you then tell the computer to FIND "suppose to" or "use to." You may also ask to REPLACE with "supposed to" and "used to" but I think it's generally better that you do each replacement manually because that's a more effective learning experience. Remember, this is not year 2000 yet, so you won't always have a computer handy.

To become a creative self-editor on the computer, you need to invent your own uses of the FIND command. If you have apostrophe problems, you obviously cannot ask for missing apostrophes, but you surely can check if the ones you did put in are proper. Since most of these find/search/replace functions can also distinguish between capital and lower-case letters, you can check if this is how you wrote "High School," College," "english," and "spanish" [all **wrong**], for example. What are **your** spelling demons? Check them out!

If you confuse "there" with "their" or with "they're," how would you employ the FIND function? How about "its" with it's"? Make your own list of what to look for. Would it help your work on sentence structure to FIND every PERIOD? [See the next chapter also!] If you come up with some ingenious editing trick with the computer's FIND action, share it with your teacher and your group mates.

EDITING WITH A GROUP MATE:

Pair up with a peer--or, when possible, with your teacher--and read your writing aloud with your partner serving as a second listener. You can do this in front of the screen or with your text on paper. It's important, however, that your helpers do not make the corrections for you; having others correct your work is of doubtful educational value. Instead they should simply stop you and point to a spot in your text and ask you if you can see or hear **any conflict. You remain in charge of your own text.**

Usually the assistance you receive while you edit aloud in conference with your teacher is extremely valuable. But don't underestimate the potential editing help available from your group mates. Try some sessions with each of them. It might turn out that one of them by temperament and language ability could provide you with exquisite coaching. And who knows what a great peer-editor you might be yourself until you offer your services. Those who help others are also helping themselves.

15 / THE PERIOD

The proper use of **the period** and of standard **verb forms.** Are the two most common difficulties in grammar for inexperienced writers. Please see the next chapter for the verbs. With periods I mean the correct marking of the **ends of sentences.** Thereby showing an awareness of sentence boundaries.

When we're speaking there's no problem. In showing the ends of sentences. Because we can help our listeners with pauses, facial expressions, and gestures. Writing is different. Here the chief aid for the readers. In spotting the ends of sentences is the period. Thus when writers leave out periods or put too many of them in their text. They confuse their readers.

The neglect or abuse of the period. Is typically of two kinds. Too many periods create **fragments** and too few create run-ons but if you do the exercises in this chapter. By correcting the two versions of this passage. You'll improve your use of the period.

?
????
????????

the proper use of **the period** and of standard **verb forms** are the two most common difficulties in grammar for inexperienced writers please see the next chapter for the verbs with periods I mean the correct marking of the **ends of sentences** thereby showing an awareness of sentence boundaries

when we're speaking there's no problem in showing the ends of sentences because we can help our listeners with pauses, facial expressions, and gestures writing is different, here the chief aid for the readers in spotting the ends of sentences is the period thus when writers leave out periods or put too many of them in their text they confuse their readers

the neglect or abuse of the period is typically of two kinds too many periods create **fragments** and too few create run-ons, but if you do the exercises in this chapter by correcting the two versions of this passage, you'll improve your use of the period.

[See the next page! ! ! ! !]

What was all that mysterious stuff on the previous page? Could you figure it out? What I did was to put in **too many periods** in the top passage, far too many, just to make a point: **FRAG**mentary sentences confuse the reader.

In the bottom passage--which is the same text punctuated differenlty--I did the reverse: I used **no periods.** The resulting **RUN**-ons are even more confusing. I call them "run-aways" because the writer runs away with the text, leaving the readers way behind. Now let's get down to business with some EDITING PRACTICE ON THE COMPUTER:

1. Enter the first of the crazy passages on your screen, exactly the way you see it on the previous page.

2. Here you have a choice: a) print out the passage with all the fragments. Or: b) take out the unnecessary periods on the screen and then do a print-out.

3. With a pen, edit or **re**-edit your print-out by reading the text ALOUD, taking out the unnecesary periods. Do not fret about commas; this is a **period** exercise.

4. If your copy turns out fairly neat and legible, that's it. But if your copy is messy, you probably want to re-edit on the screen for a second print-out (aloud).

5. Repeat #1 through #4 with the second passage, the one with all the Run-ons. But now you **insert** periods wherever you can **hear** that they belong. As you read aloud, tap your desk with your fingers at each period-pause.

6. Finally check your corrections with a group mate, with your teacher, or with the key printed upside-down below. (Don't peek too soon.)

DO NOT READ THIS KEY UNTIL YOU HAVE DONE THE TWO EXERCISES.

The proper use of **the period** and of standard **verb forms** are the two most common difficulties in grammar for inexperienced writers. Please see the next chapter for the verbs. With periods I mean the correct marking of **the ends of sentences**, thereby showing an awareness of sentence boundaries.

When we're speaking there's no problem in showing the ends of sentences because we can help our listeners with pauses, facial expressions, and gestures. Writing is different. Here the chief aid for the readers in spotting the ends of sentences is the period. Thus when writers leave out periods or put too many of them in their text they confuse their readers.

The neglect or abuse of the period is typically of two kinds. Too many periods create **fragments** and too few create run-ons, but if you do the exercises in this chapter by correcting the two versions of this passage, you'll improve your use of the period.

16 / VERB FORMS

Next to the mysterious periods, the most troublesome gremlins you'll come across in your editing work are **non-standard verb forms.** You may discover expressions like "She had went" [instead of "gone"] or "He don't" [instead of "doesn't"] or "Last night they play bingo" [instead of "played"].

One nifty way to make you more secure in the use of standard verb forms is to do exercises that allow you to play around with them, back and forth, from the plural to the singular, from the past to the present, and vice versa. Like scales on the piano or the guitar, they may not be exciting, but in the long run, the sound will get better. (Remember to read aloud as you work. Your computer will think you're chatting with her and will be appreciative.)

1. On your screen, write about a dozen sentences telling your group mates what you did during the first hour this morning. [NAME your file or document "mornpast," for instance.] Here's how one student started out:

> *I* ***yawned*** *many many times which* ***made*** *brushing my teeth easy. Then I* ***took*** *a hot shower to wake up. When I* ***asked*** *my sister to heat some water, she* ***refused.*** *I then* ***did*** *it myself. I* ***was*** *still undressed so I* ***looked*** *for some...[and so on].*

NOTE: All the verb forms are in the **past** tense. Now go ahead and write your own narrative. Remember--past verb forms.

2. Ask a group mate or your teacher to check your past verb forms.

3. Activate the COPY function in your writing program. Make a copy (second version) of your morning adventures. You could NAME it "mornpres." Put it on the screen.

4. Now make the following **changes** on the screen: A) Pretend that you do everything the same way every morning. Therefore you now put the text in what's called the habitual **present.** B) You also step outside yourself and look at yourself as if you were someone else (a "third" person), referring to yourself either as **she** or **he.** On the next page you'll see how our sample from above would how our sample from above would look:

> *He yawns many many times which makes brushing his teeth easy. Then he takes a hot shower to wake up. When he asks his sister to heat some water, she refuses. He then does it himself. He is still undressed so he looks for some... [and so on].*

Now rewrite your own text on the screen. Please remember both types of adjustment in the verb forms: the present tense and the third person singular. When you're through, get print-outs of both versions. First check them yourself and then with a group mate or your teacher.

Ω Ω Ω Ω Ω

The next exercise also deals with changing the tense and the person in the verb forms. Read the following brief passage. It's the beginning of a short story (perhaps you want to finish it).

> *Friday morning the sky was clear but dark, as if dyed indigo blue, when I woke up, a little earlier than usual. I stood on the terrace and looked out over a quiet ocean. A colibri with green pate hung by a red hibiscus flower, its wings invisible, syphoning sweetness. I had not slept well and was perspiring. The sun was already hot and looked more red than yellow.*
>
> *The breeze from the previous evening had vanished. No palm trees rustled. The bay was motionless. Funny, I thought, the radio had said nothing about a storm coming up. Then I saw something and decided to wake Wanda. I saw a hatless man in a dark suit on the beach, standing there. Just now he looked up toward me here on the terrace. His jacket bulged noticably where some men carry concealed guns....*

Now, depending on how much work you need on your verbs, choose the one of the two following processes that suits you best:

1) Enter the passage on your screen exactly as is (using the first person and the past tense). When you're done, go back and **change**

a) the "I" to "he" or "she"; and
[Use FIND & REPLACE if you like.]

b) the **past** verb forms to the **present.**

2) Make those same a) and b) changes at the very same time as you type in the passage on the screen. You modify the text as you enter it. This is a little more advanced than 1).

When you're done, print and check by reading aloud, first to yourself and then with a partner.

The modified version should begin like this: "Friday morning the sky is clear but dark, as if dyed indigo blue, when he wakes up, a little earlier than usual. He stands on the terrace and looks...."

Ω Ω Ω Ω Ω

THE HISTORICAL PRESENT:

What did you think of the practice--demonstrated in the last exercise--of writing about the past in the present? Does it make sense?

Although we can imagine the days to come and recall our yesterdays, most of us experience **the present** most vividly. How can we use this information to make our writing more vivid?

It's easy. Instead of describing an event or telling a story in the normal past tense, we fake it and pretend it's happening right now. Writers of history discovered this trick eons ago: they narrate events of the past in the present tense--as if history were taking shape right now in front of the readers' eyes. The gain in immediacy and power is startling.

You have probably done this many times though you didn't know you employed the historical present: "Last night I run into Edwin, and he is all beaten up, so I almost scream. But he is calm like an airline pilot and starts explaining ...." Do you see the trick? Instead of saying, "Last night I RAN...he WAS...I SCREAMED...he WAS calm...and STARTED" and so on, you bring the past into the instant now by manipulating your verb tenses.

"I can't do that," I hear you object. "The last time I tried, my instructor graded me down for 'inconsistent use of tense.'" Exactly. We must be consistent when we use the present--or any tense for that matter. Switching back and forth confuses our readers. Now, let me tell you why I've brought up this topic. I want you to practice "translating" a piece of writing originally created in the "normal" past tense into the historical present. This is a great exercise for all who need to strengthen their use of verb forms.

On the next page you'll find a few suggestions:

1) Dig through your folder, text, notebook, or journal until your come upon a narrative of yours (essay or short story) that you wrote in the normal past tense. Now **re-write** it, changing the verbs into the historical **present.**
If you can't find anything suitable in your own bag, move to 2) below:

2) Since most of the "Sample Student Voices" in this book are written in the normal past tense, you can turn to any one of them that you enjoyed reading and convert it into the historical present. Go down the page and change every **past** verb to the **present** like this: "At first I was scared of the computer..." becomes "At first I am scared...." and so on.

3) Write a new essay (or short story) **on the computer** in the historical **present.** Show the print-out around your group for feedback. When you've revised it on the screen, pretend that a stuffy magazine editor rejects your submission but promises to pay you $500 if you revise it by using the traditional past tense. See how easy it is to change the verb forms on the computer? Good work and good luck! Soon you'll be a wiz with verbs.

Ω Ω Ω

A final point about editing: Finding your errors is a wonderful thing--**if** you **learn** from them. The best way to profit from your discovery of mistakes is to make a record. So get your own **"Self-Editing Chart"** and make entries on it regularly (see the APPENDIX for a sample form). Depending on your access to a computer, you may wish to keep your Chart on disk. The method of keeping your own detailed notes about errors and corrections works because in the very act of making a record of your learning experience, it is powerfully reinforced.

Don't you believe it for a second!

17 / YOUR AUDIENCE: Can You Write to Santa Claus?

Why not? Many children do. He lives by the North Pole, doesn't he? But could you write **from** Santa? Let me explain.

At the end of the last century there lived an eight-year old girl in the block kitty-corner from where I now live in New York City. Her faith in Santa had been shaken by some of her little friends who claimed he didn't exist. When she turned to her father, he told her, "If you see it in the **Sun**, it's so."

You may have heard of Virginia. She wrote a letter in 1897 to the **New York Sun** and asked for the truth. And here's most of the famous reply from its editor:

> Yes, Virginia, there is a Santa Claus.
>
> Virginia, your little friends are wrong. They have been affected by the skepticism of a skeptical age. They do not believe except what they see. They think that nothing can be which is not comprehensible by their little minds. All minds, Virginia, whether they be men's or children's, are little....
>
> Yes, Virginia, there is a Santa Claus. He exists as certainly as love and generosity and devotion exist, and you know that they abound and give to your life its highest beauty and joy. Alas! How dreary would the world be if there were no Santa Claus.! It would be as dreary as if there were no Virginias. There would be no childlike faith, then, no poetry, no romance to make tolerable this existence.
>
> Not believe in Santa Claus! You might as well not believe in fairies. You might get your papa to hire men to watch in all the chimneys on Christmas Eve to catch Santa Claus but, even if they did not see Santa Claus coming down, what would that prove? Nobody sees Santa Claus, but that is no sign that there is no Santa Claus. The most real things in the world are those that neither children nor men can see. Did you ever see fairies dancing on the lawn? Of course not, but that's no proof that they are not there....
>
> No Santa Claus! Thank God he lives and he lives forever. A thousand years from now, Virginia, nay, ten times ten thousand years from now, he will continue to make glad the heart of childhood.

But the subject of this chapter isn't Santa Claus as much as **audience identification.** With that term I refer to the writer's need for a **focused awareness** of the intended recipient(s)--the "audience."

Whenever you write something--a letter, report, essay, love poem, term paper--you want **to do something to your readers.** You want your writing to affect your readers, to influence them to think, feel, or act in a certain way. For example, you may want your readers to give you an appointment for a job interview; follow travel directions; agree with you on a political issue; simply enjoy your stories; give you an A in the course; buy what you're selling; or love you and ask for marriage.

To practice this **zeroing in** on your reader, I recommend the process of sending the same letter to at lest two different people. I didn't mean that the letters would be exactly the same (only the general content), because if the audience is different, then the writing will have to be different. If you or your mates or teacher can think of assignments more appropriate to your situation than the following ones, fine, but here are my suggested topics and audiences. (Maybe we can stay with old Santa just a little longer....)

I,A. Imagine that a young cousin or nephew of yours (about 10) sends you a note asking you to explain if there's a Santa Claus or not. Write a generous and most imaginative reply (without cribbing from that editor's "Yes, Virginia" chestnut). You can make it humorous or philosophical, depending on whom you're writing to and on your inclinaton, of course. Remember that you're writing to a very young person so you must adjust your language accordingly.

I,B. Imagine that your school newspaper recently has printed an article and some letters to the editor about whether one should lie to children about things like fairies and Santa Claus. Some strident notes have been struck. You now jump into the fray with your own letter. You will strike a different tone in this letter since you're now participating in adult discourse.

Here's another topic that might be closer to home right now:

II,A. Imagine that you get a letter from a young cousin or niece (about 10) asking you whether writing on computers is all that it's cracked up to be. He or she has heard you're taking an English course with data machines and now is very eager to have it all explained in as great detail as you have time to spare. Write a helpful letter in response!

II,B. Imagine that you run into your English teacher from a few years ago in the street. When you mention the computers in your class, the teacher looks horrified and says the computers will destroy English education and dehumanize mankind. Since you don't feel that way at all, as you walk away, you imagine that you write a letter to this old-phogey yet lovable teacher explaining your experience and attitude. How will it be different from the letter your wrote to your young relative? Demonstrate it now!

Thomas Nast's Santa Claus

18 / VERSE

Auguries of Innocence

by William Blake

To see a World in a Grain of Sand
And a Heaven in a Wild Flower
Hold Infinity in the palm of your hand
And Eternity in an hour
A robin Red breast in a Cage
Puts all Heaven in a Rage

1. Copy Blake's poem in your own handwriting on the blank lines below:

----

2. Now copy Blake's poem again but on your computer screen and make a printout. How did it feel to copy it on the keyboard compared to writing it out in longhand? Do you think it's a good idea to write verse on the computer?

3. Talk with your group mates about the feelings the poem gave you. ("Augury," by the way, means "prophecy.") The poem is short, but it contains much. Can you see that while it has some big, general words, it also has some simple visual images? Making pictures is even more important when you write verse than prose. Some of these lines are easy, some are hard. Start discussing the easy parts. Then get to questions like, Why is heaven angry?

4. Compose some verse of your own now. (What's the difference between poetry and verse, if any?) Perhaps you want to try some **lyrics** and compete with Springsteen? What do you prefer to write on--pad or screen? Show drafts to mates for feedback. Don't give up until you're pleased.

Before you start, let's look at two other samples of poetry. Whereas Blake is famous and dead, these two poets are alive and not yet famous. The first one, "Pain Settles," is by Lynne Beyer:

Pain settles

in the mind's

bones: it makes

its own comfort.

Discuss in your group how pain can be comfortable? Isn't that a contradiction? Yes, poetry is the stuff of paradox. Note how Lynne's poem says so much in a few words. That's skillful compression, verbal economy.

Most English teachers by the way, write verse on the sly. And some of them are gifted poets. Ask your teacher for help with your own creation. If you hesitate because the final examination in this course is an essay and not a poem, I want to stress that all types of writing practice are useful. Verse is especially good because it helps you to listen to natural speech rhythms, to seek out strong visual pictures, and above all to experiment with language.

Finally, if you think that you're too young and inexperienced to write verse, I want you to take a look at the poem on the next page. I bet that you are older than its author.

WINTER

"Oh!" say the people in
Georgia with fright.
"It's so cold it got down
in the 30s last night."

"Ha," all the people
in Kentucky say.
"We're lucky to get it
that warm in the day.

"Your little winter is
only a splinter
compared to the cold of
a Kentucky winter."

"Haw, haw," Minnesotans say
with a jeer.
"You think that's cold?
Well come on up here.

"Your little winter is
only a splinter
compared to the cold of
a Minnesota winter."

All the Canadians
say with a smile,
"You haven't been cold
'til you've been here a while.

"Your little winter
is only a splinter
compared to the chilly
Canadian winter."

"Well," say the Eskimos
from houses of snow,
"it's warm up here,
only 30 below.

"Your litte winter is
only a splinter
compared to the freezing
Alaskan winter."

Wherever you live,
New York or Denver,
Miami or Juneau,
Always remember:

Your little winter is only a splinter
Compared to the cold of a Nuclear winter.

--Rachel Hartman

The author, who lives in Lexington, Kentucky, is 12 years old and is in the 7th grade.

Reprinted with permission from **Sojourners** [Aug./Sept. 1985.]

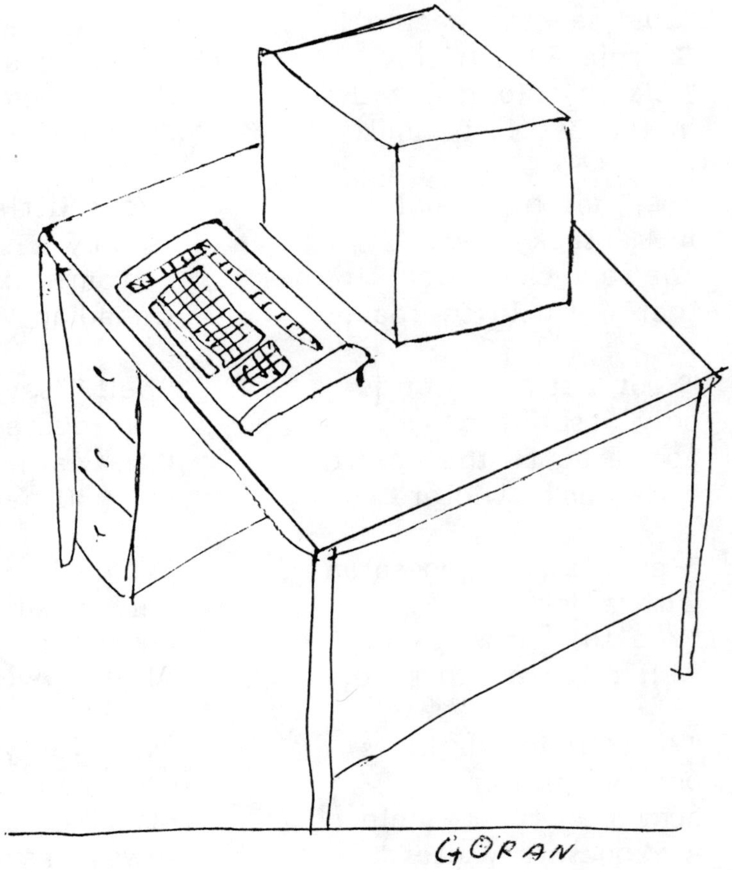

19 / WRITER'S BLOCK?

Have you ever felt paralyzed by that fear that even trying to get started on an essay assignment is an insanely hopeless idea, and that your certain failure to complete your writing on time will bring you the loss of all your friends and take you into the abyss of despair, even valley of death? Exaggeration? Well then, have you ever procrastinated about a piece of writing and been rebuked for lateness? Still doesn't apply to you? OK, then you can skip this chapter.

For those of you, however, who have been known to chew up nasty-tasting pens or to stare motionless at the computer keyboard almost forever because you know it would give you painful electric shocks if you dared touch it--for **you** this short chapter may be a lifesaver.

I used to suffer from fear of writing, but most of that is in the past. Yet I did have a touch of a block getting started writing this morning. It wasn't this chapter, though, for it is coming along easily. It was chapter 4, earlier in the book. So you can see that one action I took was not to insist on trying to write something that wasn't yet there for me. (I did chapter 4 a week later instead.) But I didn't have the time to sit idle either; I had to write something. The first thing that saved my day was recalling that I once wrote an article called "You Don't Have to Start with the First Chapter when You Write a Book." In that piece I had said that just as film-makers don't shoot their scenes in the order they appear in the final movie, so writers don't always have to do the beginning, the middle, and the end in that order. In fact, some people write the end first, and the beginning last. We do have a certain control over our own process.

"How's that point going to help me?" I can hear you complain. "I'm not a book writer, just a student trying to get through an English comp class." The answer is that the same principle applies: you don't have to write the first paragraph first. Try writing something in the middle and go from there. This approach usually gets writers warmed up and after a while inspires the wording of an opening paragraph.

Sometimes we should be grateful for a block. For we must consider the possibility that it has a beneficial function; just as a stomach-ache is a warning sign, suggesting that you adjust your eating habits, a blockage in your writing performance might be telling you to choose a **new** avenue in shaping your composition. Perhaps you don't really trust your thesis, your main point? And so your subconscious is sending warning signals pointing to health. Listen to your inner voice. In fact, this turned out to be the case with my hesitation in writing chapter 4 this morning: I later realized that my plan for it conflicted with my theory. In the end, I think, my block helped me write a better chapter.

In most cases, however, writer's block simply reflects inexperience. While seasoned authors know they can never be perfect, beginners often suffer the pangs of **perfectionism.** Give up that wish to do everything absolutely right! Practice does **not** make perfect; practice brings **progress,** and that will go a long way. Here are some excellent pointers that should help you penetrate your writing barriers:

- **Today** is the best day for writing. Begin with some **automatic writing**--it works.

- **Write every day,** even if only an entry in your log or journal. Write at least one sentence. Or a fragment.

- Set yourself a **quota** for each writing session, either a period of **time** (say 1½ hours) or a quantity of **words** (say 250 words). Remember that no one else is forcing you to write so many words in so much time--you are your own task master, so you might as well stick with it. And you'll be rewarded, too.

- Though it has been said that virtue is its own **reward,** it may be hard for you to think of a couple of pages of very rough draft as a reward. So you may want to add some self-rewards: **give yourself something when you've done your quota.** Only you know what you consider a nice reward for good work done: something to eat or to drink or perhaps a movie or time spent with a close friend. (I'm right now looking forward to my own reward: in a few minutes when I'm finished with this chapter I'll be reading **PC Magazine** and my other computer journals.)

- If you've been assigned a term paper or documented essay ("library paper"), do not wait until your research is finished. **Your research is never finished.** Only by writing will you discover what additional library work may be necessary. The time to start writing the term paper is now.

SUMMARY: Only words count. Reading and thinking--while necessary--are not writing. Though the act of writing is a process, as you've gather from this book, you also do need the end result, which is a very tangible **product.** This product comes about through the process of putting down one word after another, from the left to the right, on the page or on the screen. There's no other way, no magic (at least not until scientists have perfected a computer that can print out your spoken voice). Just as a sculptor needs a bulk of stone, marble, wood, or clay before she can shape the material, so you as a writer need a bulk of words ("rough draft") before you can shape it into a nice essay or story. Get **quantity** first; then you can seek quality. And remember, **you don't have to start with the beginning.**

In the space below I want you to make a list of the steps that **you** will take next time you feel blocked:

1.

2.

3.

4.

20 / THE PLAY IS THE THING

PRAIRIE DOG -- A Play in One Act

The cast:

MOSS = The father, in his 40s
HELEN = The Mother, in her late 30s
VAIL = The son, aged 18

THE TIME: Late February, 1910; almost dusk

THE PLACE: The Kitchen of a home on the prairie in the northern Midwest.

THE SETTING: There are two entrances to the kitchen, one leading to the outside, the other to the inner part of the house. Holding a central place to the right is a cupboard, on which, among whatever other kitchen implements are used in the play, is a large blue bowl, bottom up, a small chink off its edge. Also found in the room is a bare wooden kitchen table, three simply designed chairs placed around or near it.

AT RISE OF CURTAIN: Standing at the cupboard, HELEN is running her hand along the edge of the blue bowl. In her late thirties, the dream of green hills almost has faded from her memory--and there is the reason for this play. Seated on one of the chairs, oiling a pair of boots, is MOSS. He is in his forties. Here on the prairie he has found himself for the first time in his life. By nature, he is almost inarticulate, although bluff and good-humored. He has a tendency to appear as if he thinks with his hands.

MOSS [Singing softly virtually to himself.]: Looked at the rock to hide my face. . . Rock cried out no hiding place. . . No hiding place down here. . . .

His singing is rough and faulty, his timing none too good, spoiled somewhat by the fact that he is oiling the boot he holds in his hand to the same tempo.]

HELEN [Turns from the cupboard, takes a short step toward MOSS. Pain in her voice.]: MOSS. . . .

MOSS [Not looking up.]: Yes. . .?

HELEN: Don't sing that.

MOSS: It's a good song.

HELEN: You promised.

MOSS: Don't be foolish.

HELEN [straining.]: Please don't.

MOSS [disgruntled, but patient.]: Oh, all right. . . .

[A silence comes between them. MOSS continues to polish the boot in his hand, tunelessly whistling through his teeth, HELEN walks slowly to the door leading to the outside of the kitchen, stands there. The sound of wind is heard.]

HELEN: Moss. . . .

MOSS [The response is automatic, almost a grunt.]: Yes. . . . [He says her name almost as an afterthought.] . . .Helen.

HELEN: The wind blows.

MOSS: Don't worry. Cows are safe in the barn. Vail put them away by now. .

HELEN [Sad]: I wasn't thinking about the cows.

MOSS [Looks up. Sharp.]: What is it, Helen. . .?

HELEN: I was thinking about myself.

MOSS [A warning.] Don't' start that. . . .

HELEN: [A hand to her cheek.]: Will it be a bad blizzard?

MOSS: Don't know. . . . February is a good time of the year for a bad blizzard. . .here.

HELEN [Whistful.]: I wish we had some trees around the house. . .to keep the wind away.

MOSS: Trees don't grow proper here. You know that.

HELEN [The tone of her voice is almost that of a small child.]: Moss, dear, do you remember how the evergreens used to grow everywhere back home?

MOSS [Stiff]: This is home now, Helen.

HELEN [Slowly, as if talking to an uncomprehending parent.]: Well. . .back before we came here. Do you remember how the trees grew everywhere? And even when the snow fell you could look out the window and see the green pointing to the sky. . . . And the wind was far away, beyond the fence. . .?

MOSS: Those trees weren't ours.

HELEN [Puzzled.]: What do you mean?

MOSS: This place is ours. Even if it doesn't have any trees, it belongs to us.

HELEN: It doesn't belong to me!

MOSS: Yes, it does.

HELEN: Only you. . . .

MOSS [An end to it.]: Ours. . .Helen.

HELEN [A great finality.]: No. . .!

MOSS [Weary.]: Oh, Helen. . .let's not argue. It's too long ago now.

HELEN: It's never too long to remember.

MOSS: We've been here most ten years. You should forget.

HELEN: I don't want to forget.

[There is another silence. MOSS continues to oil the boot. After a moment, HELEN crosses to the inner part of the house. She is rubbing her shoulders.]

MOSS [Following her walk with his eyes.]: Are you cold?

HELEN: Yes, I'm going to get a sweater.

MOSS: Build more fire.

HELEN: Wouldn't do any good. That only helps when you're right by the stove.

MOSS: Well, I thought. . . .

HELEN [Is gone, on into the inner part of the house. From inside, calling.]: Where's Vail?

MOSS [Louder.]: Out patching some fence this afternoon.

HELEN: He should be back from that now.

MOSS: Maybe he is. Probably putting the cows up.

HELEN: Will he take care of the chickens, too, do you think?

MOSS: This is Saturday. Always does the chickens for you on Saturday.

HELEN [Coming back into the doorway, a sweater now thrown over her shoulders.]: He is a good boy.

[MOSS grunts noncommittally, his attention again on the boots. A scuffling is heard from beyond the door leading to the outside. MOSS Looks up.]: That's him. . .

HELEN: It would have to be him, Moss. . . There is no one else around here for miles.

[The outside door opens. VAIL comes in. His face is red from the cold. The clothes he wears are heavy and warm. VAIL is eighteen years old, tremendously interested in the growth of plants and animals he sees all around him. He is at the exact age of the most heightened awareness and his attitude reflects this.]

MOSS [Nods to Vail.]: Good evening, Son.

VAIL [Nods in return.]: Father. . . .

HELEN: Have you fed the chickens yet?

MOSS: Just going to. It's cold. I wanted to warm up first.

MOSS: Able to get the fence patched today?

VAIL: Almost. About fifty more feet.

MOSS: How's the winter wheat?

VAIL: Pretty strong. If this blizzard coming doesn't get too bad.

MOSS: Let's hope.

HELEN: How was your lunch?

VAIL [A grin.]: Cold. . . But everything is. . . I was lucky. I found a good shelter to sit under while I ate.

MOSS: What was that?

VAIL: A cottonwood tree.

MOSS: What good was that? No leaves now.

VAIL: That's what was so funny. There was leaves. That warm spell a couple of days back was just enough to bring some out. Not many. Just a few. [A laugh.] It's funny what a little change in the weather will do. That old cottonwood tree was fooled by the sun we've been having. . .and started thinking it was warm weather already.

[As VAIL has been talking, HELEN's face has been settling into grim lines of half-lost and hidden memory.]

HELEN [Very quiet, but firm.]: That was no tree. . . .

VAIL:L Sure it was, mother. It was a cottonwood tree.

HELEN: The cottonwood is not a tree.

MOSS [Warning.]: Helen.

HELEN [As if she has not heard, her voice almost a singsong. The rhythm of her words becomes like a whip]: The evergreen is a tree. . . The elm is a tree. . . The maple, the ash, the hemlock are trees. The cottonwood is not a tree. The cottonwood is nothing but a bush. A bush, a bush, a bush. . .!

MOSS [Sorrowful]: Son. . .your mother is upset again.

HELEN [Intense.] There are no trees here.

VAIL: I'm sorry. I didn't mean to. [Goes to Helen] Would you like me to finish the chickens now, Mother? [She does not reply. He turns toward the outside door.] Guess I'll go finish the chores. . . . [Walks out the door.]

HELEN [Calling, calling.]: I tell you there are no trees here. [Moves to the chair where MOSS is sitting. Behind him, she puts her hands on his shoulders.]

MOSS: Shouldn't. . . Boy doesn't know how you feel. Prairie is the only thing he knows now.

HELEN: He could remember.

MOSS: Doesn't. Never talks about the streets since we've been here. He likes it here. So do I, Helen. . . .

HELEN: I'm sorry. It's just that sometimes it's so hard.

MOSS [Trying to be reassuring.]: It'll be all right.

HELEN [Means the opposite.]: Yes. . . it's be all right. [She turns, walks toward the cupboard, where she stands looking at the blue bowl. After a long moment, very slowly.]: Did I ever tell you about this blue bowl?

MOSS: No.

HELEN: See how it's broken on the edge?

MOSS: Yes.

HELEN: Sometimes insects crawl in through there where it's broken.

MOSS: So. . .?

HELEN: They crawl in there. . .and then they can't crawl back out. I don't know why, but they can't find their way out again. And you know what happens?

MOSS: What?

HELEN: They die in there!

MOSS: Good way to catch insects.

HELEN [A sob.]: Moss. . .!

MOSS: Yes?

HELEN [Suddenly hopeless.]: Oh. . .nothing, nothing.

[Almost without warning, VAIL rushes into and through the room to the other entrance.]

VAIL [On his way through.]: I want that empty shoebox in the hallway. [Goes on out.]

MOSS [Calling.]: What happened?

VAIL [From the inner part of the house.]: I stumbled over a crippled gopher right next to the back porch. I'm going to put it in the shoebox.

MOSS: Just kill it. . . Keep it from getting at the chickens.

VAIL [On his way back through the kitchen, holding a shoebox in his hand.] Oh, Father. . .a crippled gopher can't hurt anything.

[VAIL is out the door leading to the outside. MOSS gets up from his chair, follows VAIL as far as the door.]

MOSS [By the door, looking out.]: A big one?

VAIL [A little distance away.]: No. . .just a young one.

MOSS [Comes back in. To Helen.]: Never seen anything like that boy. Acts like animals are human or something.

HELEN [Stiff.]: Close the door. It's cold.

MOSS [Closes the door.]: Sorry.

HELEN [Even more stiffly.]: Don't let him bring it in here.

MOSS: Why not? Won't hurt anything.

HELEN [Desperate.]: Don't let him. . . .

[But it's already too late, as VAIL comes back in, carefully carrying the shoebox.]

VAIL: Look at that. . .will you? Bet he tore his foot on that barbed wire by the chicken house.

MOSS [Looks in the box.]: Tore bad. . .

VAIL: I'd better bandage it.

MOSS: Don't need to be fancy. Some old dish-cloth will do.

VAIL [An unusual tenderness in his voice.]: Look at him there. . . He's so soft and helpless. [Looks at MOSS.] You hear people say how a gopher is a rat, or a dog. . .and they say gophers should be shot. Why is that?

MOSS: Eat chickens. . . That's why.

VAIL: You eat chickens! And you're wrong. . gophers don't!

MOSS: Some say that. Some say they do. Whatever, they dig holes. Horses and cows break their legs in those holes.

VAIL: You dug a hole to build this house. . .! Anyhow, a gopher is only trying to stay alive. He's got to eat. Look, animals eat plants. . .and plants eat something out of the ground. And we eat both plants and animals. What's the matter with eating something?

MOSS: Nothing. . . .

VAIL: Course not. Anyhow. . .why hurt something when it's already hurt? No reason for that. . . This gopher is hurt. Tell you what, let me nurse it back. . .then we'll let it go. If we got to kill something, seems like we can do it when they're healthy. . .don't it?

MOSS: Suppose so. [Guarded.] Better ask your mother.

VAIL (Turns to Helen.]: Is it all right, Mother? May I get him healthy before I turn him out?

HELEN [Her voice is hollow.]: No it's not all right.

VAIL [Disappointed.]: Oh, mother. . . .

HELEN [Deep hurt.]: What do you two want to do. . .? Torture me all the time?

MOSS [As if cautioning a child]: Helen. . . .

VAIL: We're not torturing you. . . .

MOSS: Can't hurt anything to let him do that.

HELEN: It can hurt me!

MOSS [Disbelief.]: What?

HELEN [Tired.]: Besides, it won't do any good to nurse the gopher.

VAIL [Protesting.]: Mother. . .it isn't hurt that bad.

HELEN: I don't care how bad it's hurt. It won't do any good. I know. I tried it once.

MOSS: You never had a gopher.

HELEN: It doesn't have to be a gopher. It's an animal. It's alive. It's hurt. . . . [Almost crying.] And it won't do any good to nurse it back. . .!

[HELEN goes swiftly to the cupboard, takes the blue bowl off the shelf. She holds it in her hands out toward the two men, a mixture of anger and fear in her attitude. She's quite apparently on the brink of a terrifying explosion of the spirit. An actress playing this part, however, must remember that the explosion cannot ever quite reach the surface. Meanwhile, MOSS stands with his feet planted well apart, his shoulders squared, his face expressionless.]

MOSS [Points at the bowl in Helen's hands.]: What's that for?

HELEN [Almost in a monotone, but building in intensity. There is lightning under every word.]: Look at this bowl! It's got a crack on its edge that lets an insect through. . .but won't let it back out again. One day last summer I took the bowl down off the shelf to clean it. Inside there was a dead spider. . .and another spider still alive. The one alive was eating the dead one. I was going to kill the spider. . .but I thought just like you did now. Vail, and I wanted to help the spider. It was only trying to stay alive. And there it was, trapped inside the bowl. . .and it was going to die. But it was still eating the other dead one. . .because it wanted to stay alive. So I thought to give it another chance, like you want to do with the gopher. And I took the spider to the kitchen door here. . .and I threw it out there on the ground. And you know what happened? [The memory is still unpleasant.] As soon as that spider that was still alive hit the ground, a big lizard jumped off the porch steps and crunched it up. And I looked at this blue bowl from which the spider came. . .and I looked up at the sky. And the sky was suddenly a big blue bowl. [Suddenly, hysterically, as if it were very dirty, she throws the bowl on the floor where it breaks.]: There. . .! The bowl is broke! No more spiders can get eaten by lizards. . .because of this bowl. [Looks directly at MOSS.]: Go out and break the sky so no more spiders can get eaten by lizards. . .because of this bowl. . . .[Looks directly at VAIL.]: Go out and break the sky. . .so that gopher won't get killed when when it's healthy again. . . .[Swiftly, now, the hysteria is gone, as well as the anger. Her posture sags. She seems afraid. Did she shiver?]

MOSS [Heavy in his voice.]: Helen. . .you been acting awful strange here recent. . . . [A bull goring.] Got to stop. . .!

HELEN [Cringing, as if his words are actual blows.]: Don't. . .Moss.

MOSS (Bulling ahead): Going to stop. . .! Going to make you. . .!

HELEN (Head down.): Please, Moss. . .don't.

MOSS: First thing is you're going to let Vail nurse that gopher back again.

VAIL (A warning.): Father. . . .

HELEN (Very tired but very strong.): You keep that gopher, you don't keep me. [As if running from a whip, an almost insane quality in her movement, HELEN rushes out the door into the freezing night.]

VAIL: You don't understand, Father. . .I understand better than you do. We got to get rid of this gopher. . . .

MOSS [Stubborn.]: No!

VAIL: Mother'll freeze out there. . . .

MOSS: Let her.

VAIL: You don't mean that. . . .

MOSS: Got to learn.

VAIL: Maybe we got to learn, too. It's better to nurse Mother back than a silly old gopher.

MOSS: Don't understand her. . . .

VAIL: I don't either. . .but we got to try. I'll take the gopher away. . . . [Goes to the door.]: I'll get her back. [He leaves. Outside the door, calling.]: Mother. . . .

HELEN [A great distance away.]: No, I won't have it. . . .

VAIL: It's all right, Mother. . .I'll take the gopher away.

[There's a long silence as MOSS stands alone in the room. He looks at the broken pieces of the blue bowl, pondering. In a few moments HELEN comes slowly back into the room.]

HELEN: Oh, Moss. . .you just don't understand, you don't understand.

MOSS (Moves to her. Puts his arms around her shoulders.]: No, I don't.

HELEN: I want to go back home where there's trees and hills. . .and not just bushes.

MOSS [Trying to comprehend.]: I can't go back, Helen.

HELEN [Sad.]: Why not?

MOSS [The words come very hard to him. He's not used to talking at such length.]: To work in a factory is no good. . .It hurt my hands all the time. Nothing belonged to me. This is mine here! Don't you remember how I used to come home. . .and I couldn't love you, I was so tired. . .?

HELEN: Moss, do you love me?

MOSS [Almost embarrassed.]: Why. . .sure.

HELEN [A sigh.]: Well. . .maybe it'll be all right.

MOSS [Reassuringly.]: Sure it will. You'll see.

VAIL [Comes back in from the outside, carries the shoebox through the other doorway, sets it on the floor the other side of the door, then turns toward his parents.]: Is Mother all right?

MOSS: Seems to be.

VAIL: Mother. . . . [His word is not a question. He says it simply to get her attention.]

HELEN: Yes?

VAIL: The gopher is dead.

HELEN: Thank you, Vail.

VAIL [Moving toward the outer doorway.]: Well, guess I'll go do the chickens now. How soon is supper?

HELEN: In a little.

VAIL: I'll be right back.

HELEN: All right.

[VAIL is gone. MOSS and HELEN break slowly apart from each other. MOSS goes back to his boots, beginning to oil one again. HELEN kneels down, starts to pick up the pieces of the broken blue bowl, then placing them in a neat pile on the kitchen table.]

MOSS [Breaking the silence.]: Helen. . . .

HELEN [From her kneeling position.]: What, dear?

MOSS [Again the words come to him with difficulty. He's not used to making offers.]: I'll plant some trees around the house. . .soon as the weather warms up.

HELEN [A small smile is seen on her face. She finishes picking up the pieces of the blue bowl. She stands up, walks to the outer doorway. Looking outward.]: Is it going to blizzard bad now?

MOSS [Squinting.]: Not as bad as it looks.

[HELEN comes back to the table. She sits down on one of the chairs, starts to hum under her breath, then the words begin to form, slowly at first, finally increasing in volume but not in tempo. As she sings, she rocks, almost as if cradling a child. But it is only a memory she cradles.]

HELEN [Singing.]: Looked at the rock to hide my face. . .Rock cried out no hiding place. . .No hiding place down here. . . .

MOSS [Soft.]: Thought you said you didn't like that song. . .

HELEN [Eyes wide and sad, but she is quieter now.]: I don't. . . .

[There is a slow curtain.]

ooOOOoo

Everybody has his or her own prairie. When were you last on your prairie? Share your feelings about the play within your group. Do you think it is **realistic** about life on the prairie? Is it meant to be? Do you identify with any of the characters? Can you see any **figurative** or **symbolic** meanings? Do you think the play also could represent family drama in other settings, like the city? Finally, start a writing project on one of the following two ideas:

1. Write a **review** of "Prairie Dog." Your teacher will tell you how.

2. Write a one-act play either by yourself or with one of your group mates. Here's one excellent way to get started:

> Sit down with one of your mates by your computer, a little to the side so that you and your friend might take turns at the keyboard. Next you **visualize** a place--any place you wish. It could be outdoors, indoors, or on a stage. In that place is a person, anyone (could be like someone you know), "A." Suddenly another character ("B") bursts on the stage and says,

B: Your father is dead.

A: [calmly]: I have no father.

Now you (and your mate) take it from there....

21 / MY WRITING PROCESS [analysis]

First you want to analyze a few examples of the kind of writing you will do in this chapter: please turn the pages and read again some of the "Sample Student Voices" between chapters 2 and 3 and 4 at the beginning of the book. You'll also find one sample essay next to this page.

If you want progress, you examine your process. In this course, this means that you survey your own past writing, giving special attention to the way you have used the computer. Your teacher will tell you whether to do the comprehensive assignment #1 below, or the more focussed #2:

1. A simple and informal method is to tell the story of your writing experience from the beginning of this course until now. If you are specific and detailed in your narrative analysis, this would be a fairly long and thorough essay.

   Start with your attitude toward computers and writing before you came to this class. Analyze your changing process and results. Respond to questions like these (and others that you make up yourself):

   Have you been drafting on the pad or the screen? Do you revise on the print-out? Do you edit both on screen and print-out? Can you give any examples of how your computer work might have improved your writing with pen? Describe your fears, joys, and hopes. Have you gone from confusion to clarity? What uncertainties or problems remain?

   What **goals** do you want to reach before the end of the term?

2. You could also focus on the way you wrote **one specific essay,** recording your tribulations as well as joys. (See Sample Student Voice on the adjacent page.) Examine the way you went about your task. Your intention is to learn about yourself: what did you do well and **what could you do better next time?**

Whether you do #1 or #2 above, at the end of your essay you want to be very specific about the **goals** you are setting for yourself and what procedures you will follow to reach them. Did you know that if your teacher requires you to reach a certain performance goal, you are not as likely to achieve it as if you yourself mastermind the goal-setting?

For more thoughts on the subject of "Self-Reliance," read the excerpts from Emerson's famous essay in the next chapter.

Stan Styk / English 090-521 / May 14, 1988 / #7

My Process: Fear of Writing

I didn't know what to expect from an English class at college. I haven't been in school since I graduated from high school seven years ago. During high school the English class didn't seem very hard. I am back now in night class trying to remember what to do to pass on to the next required English class.

When I first got into this course, I was scared because I forgot how to write well enough to pass the entrance exam. Will I ever write well enough to make a passing grade? Can I remember where to put all of the right words in a paragraph? Will my essays sound good enough to get a passing grade? The first couple of classes I felt fed up with all the essays I had to do.

Next I learned about class mode and group mode. In these modes we are either listening as a class or working as a group. I thought this was stupid until I found out it helps me with my writing. In the class mode we get help from everyone in the class room. People gave me feedback on how well or how bad I did on an essay. In the group mode I got help from my group mates. Now I have a vast amount of information to call upon every time I write an essay.

The hard part of this class was trying to write in a journal every day. It was difficult to put my feelings down in a book that a stranger may read. I felt silly writing about something that happened to me. I didn't know how other people will react to my feelings. It was hard at first to put my inner-most thoughts in a journal. I found that it was easier to write feelings down with each day that I wrote in my journal. My journal has helped me come up with four of the assigned essays I had to do for Dr. Moberg. I will continue to use a journal to keep my writing skills sharp. I don't need to fail any more entrance exams.

The professor started out slow to get everyone used to writing again. He wasn't too concerned with how we started to write. What he wanted was a lot of writing on a piece of paper in ten minutes. We would do this every time the class met. This was to improve on the amount of writing that one person can write in a given amount of time. I was told that there should be enough written down to pass the CUNY test at the end of the course. I didn't have to worry about editing until later. It took me a while to get used to writing without thinking about grammar mistakes. When I was going through school I was told to correct grammar as I wrote down my thoughts, but now I write first and then I correct my mistakes.

After the class got used to writing a lot of material in ten minutes, it was time to start working on editing. I was told to start correcting our grammar by reading the essay out aloud. I found this to be very difficult. I was always taught to read to myself and not to bother the people around me. Now I am told to read out loud so that everyone can hear what I am saying. I could not grasp this in my head until I started to find mistakes that I missed while reading silently to myself.

With only two weeks left to go I am starting to feel comfortable with my writing skills. I always find new mistakes and this keeps me from getting too cocky. The last few essays that I wrote had a lot of run-ons. I even had a fragment in the last practice test. What I do now is read everything out aloud a least five times to make sure the essay sounds right. I also listen for run-ons more than ever before because Run-ons find a way of sneaking up on me. So I am very pleased with the way the course has gone for me. It has opened my eyes to a world of writing that I didn't see during the last seven years.

Note: This student progressed well and did very well on the final assessment test.

Intermission--Sample Student Voice:

Gilbert Johnson / English 201-821 / March 27, 1986

My Computer-Writing Process

I finally produced a good paper, I think, entitled "Inner Space Before Outer Space." But even though the topic had been my choice, that essay was a rough beast to tame. I went through so much needless labor and worry before it was completed that I have made an analysis how I did this composition. I wish to avoid the same tribulations in the future.

This was not a required topic. Most of the other students wrote about Joseph Conrad's story, "The Secret Sharer." Because I insisted, I was allowed to write about the Challenger tragedy instead. The choice of topic was not the problem, however. It was that I neglected to get feedback. I wrote my essay completely alone and never showed a draft to anyone. I now understand that we don't have to write on the same topic to share drafts in the group.

I was so angry once when I couldn't get the restatement of my thesis straight in the conclusion that I cursed at the screen. My group mate sitting next to me told me to watch my language and I was embarrassed. Another time I turned off the computer in the middle of my writing. I lost my latest revisions. I now think that my difficulty all along was that I had an oversimplified view of this enormous space accident. For some time I refused to change my draft along with the changes in my head.

I also ran into computer problems. It wasn't the computer's fault. It was mine. Though I am not a fast typist and normally write the first draft with pen, this time I did everything on the computer, from beginning to end. It was like a need to show I could do it but it was too early. I guess because I wrote on a different topic, I felt like a lone cowboy and wanted to do it my way. I now know that my best process at this time is to write the first copy on paper and then move to the computer.

In a sense I also had a printing difficulty. I must have reprinted my essay more than twelve times. I want to change that part of my process. For example, instead of making just a few changes on the screen and then rush to the printer, in the future I would like to sleep on the draft for a couple of days and then revise thoroughly, first with pen and then on the screen before I print again.

When I got the essay back from the teacher, however, I was not disappointed in the grade. All the work had paid off. My teacher told me that I should not be so hard on myself. So in the end I learned about how to improve my writing process from all the frustrations of this assignment. I don't like to admit that progress sometimes costs pain, but I guess it is true.

Intermission -- Sample Student Voice:

Alex Augustus / English 090-084 / December 1, 1988 / #8

Teenage Freedom

I must admit that teenagers do have more freedom and power of adults without adult responsibilities. Still, I don't think that the teenage years are the happiest times of a person's life. As a teenager I should know. One of the biggest problems is peer pressure. I see stories about it on the news, and I see it all the around me. The pressure is not just from our peers but from everyone and everything around us. When it comes to peer pressure, we, the teenagers are faced with difficult decisions. We're pressured into drinking, sex, and taking drugs. The biggest reason is that we want to be accepted, liked, and part of the in-crowd. We are also pressured by our parents. They are always on my case about doing well in school and getting high grades. This causes a mental strain because we're scared to let our parents down.

Maybe it'll be best if I speak from my point of view. I'm the type of person who likes to be well known. That's one reason why I joined the football team in high school. From that I knew a lot of people, and a lot of people knew me. You could have walked into that school and asked almost anyone whether or not they knew me. If they said "little Alex" that was me. Afterwards I began to see people drinking heavily at parties, smoking and so on. Guys on the team were taking mega-packs, which consisted of five different pills that were used to "bulk up" while we trained and lifted weights during the off season.

Almost everyone was pressured into sex and relationships (I was). I know four girls personally who had kids before we graduated. That's not to mention the ones I knew who had abortions. When it came to grades, I was more scared and nervous of what my parents would say and their reaction to it, more than the grade itself. They were constantly yelling at me about doing good and getting high grades so I could get into college. It got to the point where my mother blamed everything on football, and as a result I was forced to quit the team. It became so bad till I used to look forward to going to work to "escape" from it all. How ironic! Most people can't wait to leave theirs, and I couldn't wait to get to mine. Even now I wish I could just get up and go to work. I wouldn't have to worry about homework, tests, midterms, finals, peer pressure, or parent pressure. I'd just go to work, put in my eight hours then go home.

Also a majority of the teenagers have some sort of bill to pay. I know I do. I have to pay my credit card bill, as well as book fees, transportation, lunch money, clothes, and so forth. When all the pressures build up, some kids can't handle it and just commit suicide. Then when one teen does it, someone else with similar problems commits suicide as well.I made a vow to myself that no matter what happens or how bad things get I will never kill myself. I love myself too much for that. That's why I must conclude that though the teenage years might contain some happy times, they can also be the most trying times in life for most people.

22 / PERSUASIVE ARGUMENTATION

Human beings have argued among themselves since the beginning of time. One subject they still argue about is which is the most persuasive way to argue. I propose that, in many writing situations, the traditional pattern for an argumentative essay--which goes back to ancient Greek oratory--is not always as persuasive as it could be.

Before I explain myself, you should know that this is a book about the basics of effective writing and not a course in analytical thinking or advanced English composition. So what follows below will not offer instruction in logic and the rules of evidence but will give you a simple pattern for writing persuasively.

I basically go along with of the traditional model for the argumentative pattern of development. Here's a simplified list of the usual steps:

1. Introduce your position, thesis, or main point.

2. List several **reasons** why the readers should believe something or act in a certain way.

3. Somewhere along the way you also anticipate your opponents' objections and try to refute them.

4. Quickly restate the gist of your chief argument and urge your readers to agree with you.

It all sounds very reasonable. In reality, however, people are not usually swayed by sheer reasonableness. I want to focus on item #2 and its key term, **reason.**

Modern psychology has taught us that most people resist "reasons" why they should change their actions or opinions. They tend to look beyond the glittering logic and facts and ask, "What's in it for you? How come **you** are pushing this?" I say, then, if that's what most people want, give it to them. The job of writers is to be of service to their readers. Instead of giving **reasons** why the readers **should** accept this or that [do not **should** on people--they do not like it], I prefer to give the **reasons why I hold the position I do.** The difference may seem slight, but it is subtly powerful. Here's an example.

Let's imagine that it's time for the election of student officers at your school. You're writing an article for the campus newspaper in favor of Ms. Candidata. Now what I'm proposing is simply that instead of writing, "Here are three good **reasons why you should** vote for her," I would write, "I want to give you three good **reasons why I have come to believe** that Ms. Candidata is the best choice." In the second version it is still obvious that I, the writer, am advocating the election of Ms. Candidata. But I'm not intruding on the readers by telling them their business; instead, I'm sharing with them some of my own business. Most people appreciate that and listen far more carefully than if they felt someone was trying to tell them what to do or believe.

The approach I advocate is less formal but takes more courage because you are self-revealing. And that's a quality you'll find in most good writing. Let's look at the issue of courage for a moment.

When you want to convince your readers of the merits of your position on a controversial issue, it would help you to have the courage to take full responsibility for your viewpoint. Your readers will respect you if you strike a initial note somewhat in this manner: "This is my opinion and here are my reasons for them." When you've finished developing your argument, you may or may not have convinced your readers, but your forthright approach will have promoted good will and communication and will persuade more strongly than if you had been trying to win debating points through "objective facts and eternal truths."

How would you react, for instance, if you were reading an essay that sounded like this: "The scientific position taken in this article is of course the only one acceptable to intelligent people. The arguments to be presented in due course are irrefutable and logical as will be proved beyond a doubt...." Wouldn't you be offended? Yet through the years I have seen numerous writers open their essays in that manner, hiding behind impersonal, passive constructions and pompous allusions to reason. Even if the reasoning and the facts are impeccable (which they often are not), the arrogant tone puts the readers on the defensive.

My proposed modification of the usual pattern for persuasive writing--focusing on the **reasons for the author's views**, rather than the reasons why the reader should adopt them--has worked for me. More importantly, I've been impressed by the writings of those who develop their essays in that manner. Although there are some occasions when this approach is unsuitable (most scientific, technical, or business reports, for example), I believe that it will work for you on most occasions in school and beyond.

Try my pattern when you write about Emerson's essay on the next page. You may also want to read the Sample Student Voice on p. 141 in the Appendix, "Money Isn't as Important as You Think."

Self-Reliance

Excerpts from Ralph Waldo Emerson's Essay

What I must do is all that concerns me, not what people think. This rule, equally arduous in actual and in intellectual life, may serve for the whole distinction between greatness and meanness. It is the harder, because you will always find those who think they know your duty better than you know it. It is easy in the world to live after the world's opinion; it is as easy in solitude to live after our own; but the great man is he who in the midst of the crowd keeps with perfect sweetness the independence of solitude.

For nonconformity the world whips you with its displeasure. And therefore a man must know how to estimate a sour face. The bystanders look askance on him in the public street or in the friend's parlour. If this aversion had its origin in contempt and resistance like his own, he might well go home with a sad contenance; but the sour face of the multitude, like their sweet faces, have no deep cause, but are put on and off as the wind blows and a newspaper directs....

The other terror that scares us from self-trust is our consistency; a reverence for our past act or word, because the eyes of others have no other data for computing our orbit than our past acts, and we are loath to disappoint them.

A foolish consistency is the hobgoblin of little minds, adored by little statesmen and philsophers and divines. With consistency a great soul has simply nothing to do. He may as well concern himself with his shadow on the wall. Speak what you think now in hard words, and to-morrow speak what to-morrow thinks in hard words again, though it contradict everything you said to-day. -- "Ah, so you shall be sure to be misunderstood." -- Is it so bad, then, to be misunderstood? Pythagoras was misundertood, and Socrates, and Jesus, and Luther, and Copernicus, and Galileo, and Newton, and every pure and wise spirit that ever took flesh. To be great is to be misunderstood....

I hope in these days we have heard the last of conformity and consistency. Let the words be gazetted and ridiculed henceforward. Instead of the gong for dinner, let us hear a whistle from the Spartan fife. Let us never bow and apologize more. A great man is coming to eat at my house. I do not wish to please him; I wish that he should wish to please me....

Insist on yourself; never imitate....

Ralph Waldo Emerson's ideas about proper personal conduct may appear either conventional or provocative depending on one's own views. If, after reading the excerpts from "Self-Reliance" on the previous page, you find his thoughts intriguing, you may want to go to the library and read the whole essay. Ask your teacher. Now, proceed with the steps below in completing the writing assignment in this chapter:

1)The first task is to **paraphrase** the essay: what are Emerson's two or three main ideas in this passage? Write them down on your pad. Go on to #2:

2) Next, your whole group discusses these main points with the aim of extracting a generalized sentence that could serve as a **thesis statement.** If you can all agree on one **main point,** fine. If not, stick to your own thesis.

3) Continue the group discussion around the questions below. Keep taking notes. Later they will be helpful in writing your draft.

a) Emerson wrote his famous essay a century and a half ago. How did he know about data processing and computers way back then? (See the words "data for computing" in the third paragraph.)

b) How do you feel about the style, vocabulary, and spelling? How is the language different from today's English?

c) How well does "collaborative learning" mesh with "self-reliance"?

d) Explain to your group or class mates if Emerson's ideas seem wise, crazy, out-of-date, modern--or what? If you want to say that you have "mixed feelings"--that Emerson's opinions are partly sound and partly unsound--you must explain very specifically to your mates exactly which ideas you feel which way about. In the end, however, it is best if you can avoid fence-sitting. It may in fact mask indecisiveness--and lack of self-reliance. Instead take a firm stand either for or against Emerson's basic thinking in "Self-Reliance."

4. Your **writing assignment** in this chapter is to either favor or oppose Emerson's ideas in "Self-Reliance" in an essay of your own. You could think of it as a **response** to him, perhaps in the form of a letter. Give the **reasons** that brought you to your views and support your position thoroughly with several **examples** from your own readings and experiences.

ooOOOoo

Follow the same steps as above in writing another essay in which you respond to the "letter to the editor" of the **New York Times** that you see on the next page. Argue as persuasively as you can either for or against the views of the letter-writer.

THE NEW YORK TIMES, FRIDAY, JANUARY 10, 1986

Letters

Schools Ask the Wrong Questions

To the Editor:

Fred Hechinger (column, Science Times, Dec. 31) says that despite many calls for reform, "There has been no serious departure from the way the American school has been organized for more than 100 years." Why such resistance to change?

One answer may be that the testing system is the tail that wags the dog; we may need to trade in the old tests for new ones. Teachers are constantly training children to pass tests that certify them for passing into next higher levels. These tests require manipulating word and number symbols or regurgitating knowledge such as the difference between a paramecium and a euglena. The tests demand that the old system remain intact.

What old system? The system that classifies knowledge into such subjects as English, history, biology and math — the system that pins students to desks and forces them to swallow masses of material of little interest to them. Forty percent drop out of high school before graduating. Isn't there a strong message in that statistic?

What if local and national tests addressed such questions as: How can people reduce the spread of venereal and respiratory diseases? How can adequate nutrition be maintained on a modest budget? How can children be disciplined without battering? How can conflicts be resolved without violence? How can automobile safety be increased?

We die by the millions for lack of education in these subjects. How many go to school and never learn how to prevent a pregnancy, cover a cough, balance a checkbook or boil unsafe water for the baby?

Of course, both biology and social science would have something to contribute to disease prevention. History and English literature would have something to contribute to war prevention. But emphases would change. New and different knowledge would have to be sought to answer the questions. Reading and writing skills would bloom if students feel the subject matter is vital to them.

If schools are to change, we will have to change our notions about what constitutes basic subject matter. Educators with powerful interests in the old ways of classifying knowledge will not be able to change. We, the citizens, will have to demand new kinds of tests. By asking new questions, we

If we had tests that measured student skills in citizenry, i.e., mutual care and decent management of human problems, we would find the schools teaching to these tests. Instead of demanding time for their pet subjects, teachers would have to pool their knowledge and organize it so as to encourage enlightenment on major human issues.

will see educators reorganizing efforts to help students master new tests. The new tail might then wag the new dog.

RACHEL M. LAUER
New York, Dec. 31, 1985

The writer, former chief psychologist of New York City schools, is director of Pace University's Thinking and Learning Center.

Intermission -- Sample Student Voice:

Chitreka Singh / English 090-511 / November, 30, 1987 / #9

A LETTER TO EMERSON

Dear Emerson,

It is the most fascinating essay I have ever read. I interpret it as a mixture of profound prose and poetry. Once I understood the essay clearly I found I relate to it. Your style of writing is realistic. "Self Reliance" is a wonderful piece of writing. Emerson, I am totally in favor of your ideas and opinions about self-reliance.

When I first read the essay I was confused. I was mainly confused about "data for computing" in paragraph 3. I thought of a computer, but this essay was written in the last century before computers were invented. Anyway, I reread the essay and our group discussed it. We shared our ideas by reading each line and giving our best explanation as we read. Sharing is a wonderful thing. After we exchanged views, I learned that data is information and computing is the brain or memory bank. I then combined my explanations with my group mates' and I found your essay very interesting and practical.

Here is one example of why I am in total agreement with your philosophy of "Self Reliance," namely in the words "one must insist on being oneself." Growing up as a child I was obligated to my parents, brothers, sisters, relatives and older people in my society. It did not matter what I thought or said, because in their opinion they were always right. In the past I tried very hard to please people. I tried to do things that would make other people satisfied and happy.

I remember vividly a particular woman who was always trying to live my life for me. She used to dictate the way I dressed. She was very controlling and possessive. Whatever she said had to be acknowledged. No one had the nerve to disregard her command. She more or less ruled with an iron rod. She strongly disliked me simply because I never let her to get to me.

Another reason why I am in favor of your ideas of "Self-Reliance" is what you stated in paragraph #2: "For nonconformity the world whips you with its displeasure." Yes, I discovered that a long time ago. Once I am obedient to someone, I am deemed a nice person. The moment I decide to do what I think is right, I am tagged a disrespectful and selfish person. That is why I find your essay so realistic. It applies to today's style of living.

Here is yet another reason why I admire your writing. It is stated in paragraph #3: "The other terror that scares us from self-trust is our consistency." Sometimes I get scared of being inconsistent, of contradicting myself. Will I let my opinions waver? Will I bow to a person's every wish and command? The fear of inconsistency corrodes my mind so strongly that I am afraid that the world might see my past acts that I have been concealing for a long time.

In paragraph #4 you mentioned statesmen (politicians), philosophers (tutors), divines (religious leaders). Great leaders like Jesus, Luther and others were misunderstood. Those were men who had great humility, intelligence and sense of purpose. People ridiculed Jesus because He stood for truth and justice. He is still being ridiculed today. (I believe He is alive). King was a just man with a dream to free Black people. He shared his dream with other people, but his life was cut short because there were people who did not like what he was doing. Those are only two examples. I could give you a few more but time is limited. When I read about political and religious leaders who were misjudged and misunderstood because of their concepts of life, I can agree with your statement "TO BE GREAT IS TO BE MISUNDERSTOOD...." Sometimes I'm misunderstood, too, so maybe I will become great.

Emerson, there are so many more reasons I could give you why I admire your style of writing and your ideas and opinions that it would probably take a very long time to pen them. I don't want to sound repetitious, but your essay really enlightened me. It helped me to respect myself, to think about myself. I no longer believe that a gullible person is a humble person. I compromised my convictions and opinions just to make other people happy, but I only succeeded in making myself miserable and confused. I don't want you to think I am a cold-hearted person. Actually I am born-again Christian and it is my duty to make people happy and joyful, which I do. But I still take a firm stand when necessary.

Intermission -- Sample Student Voice

Miguel Reyes and Eugenia Cartagena / English 090-084 / December 5, 1988 / #9

In Response to Emerson's "Self-Reliance"

Emerson's main point in "Self-Reliance" is, as he said, "Insist on yourself never imitate", meaning that one should be self-reliant, that one individual should do what he or she wants, not what others think should be done. Emerson said that anyone who is different and thinks differently may be misunderstood. To be great is to be misunderstood. Wouldn't anyone agree? We--Eugenia and Miguel--have argued and reread the essay over and over again. Despite all the long words and the style of his writing, we have basically come to understand Emerson's philosophy. Emerson must have been great, for he was really misunderstood. Such great individuals as Socrates, Jesus, Luther, Copernicus, Galileo, and Newton were misunderstood. Who knows--maybe even Miguel and I?

To further explain why we I agreed with Emerson, we want to give some examples. Many young adults today experience peer-pressure. They are influenced into acting certain ways and doing certain things. This idea leads us back to Emerson's essay. Why can't adults be more self-reliant? People tend to please others instead pleasing themselves first. Why is it that the environment is so full of followers? This now brings us to the real reasons why we believe in Emerson's philosophy.

Miguel: He has come to believe that people should please themselves before pleasing others. He feels that anyone who acts as an individual is considered one. One that follows is merely just that, a follower. Which would someone rather be, a chief or an Indian? Miguel thinks that he would rather be heard instead of he having to listen. He would rather be followed instead of doing the following. Even though he does admit trying to please others, he tries very hard to please himself before anyone. It all goes back to Emerson's essay, "Insist on yourself, never imitate."

Genie: She feels a person should not depend on others and not "copy" each other. Many people feel that if they follow the crowd they will be noticed. In her opinion, a person who follows the crowd will totally go unnoticed. She has gone with the crowd a ew times. When she was in high school, she saw practically everyone smoking. She decided to try it because she didn't want to be considered a "party-pooper." The result of going with the crowd was being "grounded" for a month.

To get back to some specifics in Emerson's essay, we would now like to focus on this remark, written some 150 years ago: "No other data for computing." How did Emerson know about data processing and computers? Right away we discovered in the dictionary that "data" means "information organized for analysis or decision making," and "computing" means, "to determine by numerical methods (calculate, reckon, count)." So as a reader can really focus on these words they won't be misled. This means that words have many different meanings and they could have been otherwise different a century and a half ago.

Before concluding, we want to elaborate on "collaborative learning" and collaborative writing. While we were writing this essay as a joint-partnership, we ran into a difference of opinion. This goes back to Emerson's self-reliance. We had an argument over a part of the essay where we wrote our own opinions. We finally realized that there is quite a bit of self-reliance in each of us. We--Miguel and Eugenia--wouldn't go back to try to please the other no matter what. We can actually say that we understand better what we went through, and what Emerson went through.

Don't forget, "Insist on yourself; never imitate...."

NOTES

Appendix

Sometimes I put messy thoughts and feelings in my journal.

TALKING AND WRITING ABOUT A NOVEL

<> <> <> <> <> <> <> <> <> <> <> <> <> <> <> <> <>

Most of you have picked up a novel now and then just to read for the fun of it. Reading a novel in English class can be done "for fun," too. But for class you don't read in solitude; you share your experience with your group mates and class mates in plenty of discourse--talking and writing about the book. This makes for more thorough analysis of our reading experience than when we read alone. Here's an outline for the process of reading/writing a novel in a writing class. Your teacher might ask you to follow some of the steps in this procedure:

DOUBLE-ENTRY JOURNAL-KEEPING

Can a human being eat an elephant? Sure--one mouth-full at a time.... Some novels are pretty thick, so it wouldn't be good to read the whole thing without first discussing its parts in class. Here's an example. Michael Dorris's splendid novel, *Yellow Raft in Blue Water,* naturally falls into three parts: the stories of 1) Rayona, 2) Christine, and 3) Ida. When I assign this novel, I ask my students to take one week for each of the parts. For each of the three readings, they follow these steps:

A) ***Read*** the section in the novel with a PEN in hand, making plenty of markings, marginal notes, underlinings, and other comments. I would consider it penny-wise and dollar-foolish to leave your book untouched by human pen--just to get a few extra pennies when selling it back to the store at the end of the term. Marking up a textbook is a good technique for learning.

B) ***Journal Summary***: After the first reading, open both your ***journal*** and the novel so you can use them simultaneously. As you read through the text again slowly, enter in your journal in your own words a ***summary***--or "paraphrase"--of what happens in the story. Just imagine that someone asks you, "What goes on in this book?" and you respond clearly but briefly. You can put down some of your interpretations and feelings at this time if you wish, but you can leave most of them for the next step.

C) ***Journal Response***: Now comes the exciting part of this reading/writing process: you are going to have a dialogue with yourself. At this stage you ***re-read*** your own summarizing notes in your ***journal*** and then add comments on your own comments by elaborating and adding thoughts, feelings--above all ***questions***--memories, ears, joys that occur to you in connection with your reading experience. Be specific and use page references to the novel (this will be helpful in group discussion later).

[continued on next page]

NOTE: During D) and E) below, each group has a ***reporter*** who keeps a record of the proceedings, either in pen or on the computer, if convenient.

D) ***Group Discussion*** in class: 1) One of the group members reads a ***summary*** aloud from her journal. 2) After that, another member reads his ***response*** aloud to the group. 3) Next, the whole group compares notes by exploring different interpretations and opinions.

You don't have to come to an agreement about the meanings in the novel, but at least agree on what you disagree about. The group reporter keeps careful notes.

NOTE: If the class has access to the Realtime Writer program, the next step could be done as an RTW "conversation."

E) ***Role-Playing***: Most students find this to be the most rewarding step in the novel-reading process. Each one of you in the group will play the part of one of the major or minor characters in the novel--plus the author. For example, if you're studying *A Yellow Raft in Blue Water*, some of your choices would be these:

THE WOMEN: Rayona; Christine; Ida; Clara; others

THE MEN: Micahel Dorris (the author); Lee; Elgin; Fr. Hurlburt; others

You'll often find that some of you like the novel, and some don't. It would be natural that the group member who chooses to play the author is one who likes the book. Please note, too, that you don't at all have to choose a character of your own sex; in fact, often it can be more interesting to play the part of someone not your own sex.

One way to go in this role-playing session is to select pivotal scenes in the novel and ask each other why "YOU" acted and spoke the way you did, and why the author wrote it that way. Remember, you don't have to solve all the questions--the important thing is to ***raise*** questions.

NOTE: If the class has access to the Realtime Writer program, this step, too, could be done as an RTW "conversation."

F) ***Class Discussion***: Finally each group reporter in turn reads the group report on the novel discussion/role-playing to the whole class so that you can find out what all the others are thinking and writing about this novel. Your teacher will co-ordinate this step.

G) Usually in the end you write your own essay about the novel or about a topic that emerged from your experience in reading it and talking and writing about it.

Or you might want to write your own novel. If that sounds like eating an elephant--how about writing a short story?

<> <> <> <> <> <> <>

Using the REALTIME WRITER Program

##

More and more networked computer labs are now installing Realtime Writer (RTW for short) for English and other classes. This is a magnificent program that allows the students to actually hold a genuine "conversation" IN WRITING: you enter a silent world of textual discourse--no speech--all writing and reading. This artificial world is a literacy lab. Here's one way to get started:

To change from your regular word processor to RTW:

1. Quit the other program to get a C:> prompt.

2. Change directory: Type CD\RTW. (Use the BACKslash!)

3. Type RTWLOG IN. Then press ENTER.

4. When asked about class ID, type class ID. Press ENTER.

5. When asked about your own ID--type your FIRST NAME. ENTER.

6. When asked about your PASSWORD, repeat FIRST NAME. ENTER.

7. When asked about reason for joining, press ENTER.

8. When asked about comment--press ENTER.

9. Now, for the HOT-KEY: hold down the Alt key and press F9.

A. You're now ready to write in the private "Compose Window."

B. When you want to send your comment to the top public "Dialog" window, press F10. All this is taking place in Channel ONE, which is the whole class.

C. If you want to join just your own group, go to your group Channel (your teacher will give you a number), press Alt-C, and then the number of your Channel. Press ENTER.

D. When you want to read all the comments in the top Dialog window and need to scroll, hit the ScrollLock key. Once your cursor is in the top window, scroll as usual by using the UP and DOWN arrow keys.

E. When you're ready to quit Real Time Writer (RTW), press Alt-E (E stands for "Exit"); then type RTWLOG OUT, plus ENTER.

Your teacher will undoubtedly tell you what to do when you are in the world of Realtime Writer, yet many teachers also welcome student suggestions. Why don't you take a look below, discuss some of these activities in your group, and then let your teacher know of your group's preferences.

@ Early in the term: Introduce yourselves all around in your group so that you'll get to know your mates.

@ Hold more formal INTERVIEWS for your group essay (Chapt. 9) either with one person or with the whole group.

@ Bring up questions about the day's homework within your group in preparation for class-wide discussion.

@ Discuss a specific GRAMMAR problem. No--don't moan, for though this topic often seems unexciting, many students have learned much from discussing grammar on the RTW screen.

@ Instead of reading aloud to your group mates from your JOURNAL, you could copy some highlights onto the RTW screen--asking for feedback.

@ If you're reading a novel in class, you might play-act different characters from the book, engaging in an imaginary dialogue on the RTW scren.

@ The first time or so you'll probably just spontaneously enjoy the excitement of "talking" with your keyboard and screen, but as you get more accustomed to this new medium, you might want to consider some of these pointers:

1. Do not rush. In the beginning, revise and edit your words a little before kicking them upstairs to the dialog window.
2. If you don't understand somebody else's contribution, ask for clarification.
3. Try to write some sentences that are longer than 3-4 words.
4. If you feel ignored in the "conversation," ask for responses. You might also ask WHY nobody replied.
5. Avoid sarcasm, vulgarities, and other putdowns.
6. If you are in your group members are all on the same RTW channel and you feel that too many ideas are jumping all over the screen, suggest that you pair off on different channels to keep the conversation more focused.
7. If you feel your RTW "talk" for the day was particularly interesting, ask your teacher for a print-out so you can take it home and study for further discussion.

PARAGRAPH ORGANIZATION

[1] One of the keys to effective paragraphing is the division of essays and other pieces of writing into groups of sentences called paragraphs.

[2] Although a paragraph can be quite short and consist of only one sentence [as in par. 1 above] or could be very long with a dozen or more sentences, the typical paragraph in our time contains five or six sentences like this one. The suggestion about length is that each paragraph should be long enough to explain its topic or main point. It should not be so long, however, that it digresses into a discussion of topics that deserve their own treatment in other paragraphs. Since the topic of this paragraph is length, it would not be a good idea to talk about the purpose of paragraphing without starting a new paragraph.

[3] Your purpose in dividing your essay into the sections we call paragraphs is to help your readers see how you've organized your writing. Writers normally discuss one point or topic at a time, showing the progress from one to the next by indenting an inch or so for new paragraphs. All good writing is organized. Paragraphing reflects organization.

[4] One chief quality of a good paragraph is unity. By this term I mean that the author sticks to one topic. In fact, the word unity means one-ness. Some writers try consciously to create a topic sentence which states the main point of each paragraph, but I consider this optional. In the paragraph you are now reading, the first sentence serves as a topic sentence.

[5] Another thing effective writers do is mind their transitions between paragraphs. In the previous sentence, "Another" is a transitional tool that ties paragraphs 4 and 5 together. For other transitional words and phrases see the chart in the APPENDIX that lists **Conjunctions.**

[6] Finally, good essays often end with a paragraph that summarizes or restates the thesis (main point) of the whole piece of writing. Thus one could say that a typical, good paragraph, which could be of varying but usually medium length, aims to show organization of ideas, has unity, and flows smoothly thanks to transitional devices. Effective paragraphing leads to the effective clarity that you want in your writing.

CONJUNCTIONS--Handy Reference Chart for Joining Clauses (& Linking Sentences):

ΩΩΩΩΩΩΩΩΩΩΩΩ

I. To join clauses within a sentence, use CONJUNCTIONS.
Here are the most common COORDINATING conjunctions.
They introduce independent (or "main") clauses:

AND, BUT, OR (& NOR), FOR, SO, STILL, & YET.

B. And here are some common SUBORDINATING conjunctions.
They introduce dependent (or subordinate) clauses:

				LIST OTHER SUBORDINATING CONJUNCTIONS:
after as before since until when while	[time]	as because since	[cause & effect]	______ ______ ______
although as if if though unless whereas whether while	[comparison, condition, etc.]	that which who	[relative pronouns serving the same function as subordinating conjunctions]	______ ______ ______ ______

II. Use ADVERBS or adverbial phrases to link sentences together within paragraphs or between paragraphs. They are used for transition, to give coherence and a smooth flow to the development of your writing.

Remember, however, that these transitional devices link sentences and DO NOT JOIN CLAUSES. Here are some common adverbs:

			LIST OTHER ADVERBS:
first	in fact	however	
second	of course	nevertheless	______
third	as a result	on the other hand	
in addition	consequently	otherwise	______
further	therefore		
finally	thus	______	______
______	______	______	______
______	______	______	______
______	______	______	______
______	______	______	______

MY SELF-EDITING CHART

THE WAY I WROTE IT:	EDITED FOR STANDARD AMERICAN:	WHAT I WANT TO REMEMBER:
[2 samples--then you do your own!]		
(1) my sisters lover	...my sister's lover	...I need an apostrophe (') to show ownership, also called "the possessive."
(2) I did the best to keep the marriage together I tried everything.	I did my best to keep the marriage together. I tried everything.	If I read it aloud, I can <u>hear</u> the definite PAUSE after the word "together," and then I know I have to put a period there.

[Suitable for photocopying]

Remembrances

VOLUME 1 NO. 1 NOVEMBER, 1988

Have You Ever Missed Someone?

Grandma: My Very Best Friend

by Jamal Mindod

"Have you ever missed someone?" This is a question that had to be asked by someone to almost everyone at one point in life. Someone students and I would like to express our feelings of remembrance towards someone we were separated from.

The first time I asked myself this question, I had to think about all the people I was separated from in life. Then I realized, the first time I asked myself this question was when my grandmother died.

I did not know my grandmother for a long time because I was only four years old when she passed away. But I do remember the good moments we had together, they still make me feel good.

These same good moments that I remember is what makes me miss her, and want to recapture the moments we had. But I know it is impossible because she is in one world, and I am in another.

But missing a person is something that everyone is bound to face in life. The way a person is affected by the person separated from them, will depend a lot on what kind of relationship they had.

Missing Thoughts

by Allison McDonald

I lost someone that was very dear to me. How I miss him. It doesn't seem that long ago that we were together. It hurt so much. The mention of his name brings pain to my heart. It seem where ever I go, I see his image. He was five feet seven inches tall, had a carmel broown complexion, black hair and was very well mannered.

I miss the wonderful times that we shared. There's no other like him. We shared days, nights, problems and laughter with each other. We were insparable. No day would pass without us seeing or speaking to each other.

The love we had for each other seemed endless. I believed he was the only one for me. I couldn't see myself having a life without him, not bieng part of his we had so many plans for the future that were destroyed. The true colors of our love had been shown. Agony and tears have been shared. Our laughter turned into agreements and distrust.

I will always miss him, because of all the things that we shared, together. The times that we spent together were very nice, Even though the relationship didn't work out. There will always be a place in my heart for him, because I learned a great deal during the course of the relationship.

How I live for the day I hold you in my arms again.Never will I let you go. I miss your smell of flowers, your embracing kisses and your forever pretty brown eyes. More and more I'm in need for your touch. That smile which no one can copy, and those beautiful noises that came from your lips. It's just the little things you did that made my days worth living.

Oh lovely one, give me back your sight. Give me back the food that feeds my love towards you. All I ask is for one day. One day of you, so life's eternal vows can be taken.

I remember taking you to the airport, but I never left. Part of me was taken away, and a little was kept. I cried all the way home until I ran out of tears. Come back and let me have what was once mine. The part of me that was taken away...Porsche

by Arden Als

A Special Aunt

by Kim Kearns

Many people have missed people they loved. I especially miss my aunt. A heavy set fair skined lady from Jamaica. She was kind and generous. She was always there when the family needed her. She used to live with us, so we were very close.

I loved my aunt like i love my mother. When my brother and I were younger, she was our nanny. She taught us academic things before we went to school. We were advanced students in away, and were ahead of everyone. Even when we would come home in the afternoon, we would get right to the books. She made learning extremely fun. She made my brother so smart that the counselor was willing to skip my brother a grade. My father was pleased but had refuse because my brother would not adjust to the upper classmen.

My aunt was upset because she wanted my brother and I to have the best schooling.

One night my mother received a phone call. It's was about my aunt, she had a heart attack and passed away. It was the most shocking thing I ever experienced. My mother was in shock because that was her sister. During that time she was married and living in Yonkers,so it was kind of hard getting up there at 3 a.m. We made it a must to go up there to see her before the coroner came and took her away.

I was so shocked . All our dreams were shattered. My mother blamed her niece for my aunt's death, and my mother cursed her and said that,'' she will suffer for the rest of her life''.

The day she was buried was also my birthday. That was the worst day of my life. I had a funny feeling in my heart that if I would turn around, I would see her calling me from behind a tree. I knew it would never happen. The thoughts of her are always constant and I will cherish her memory forever.

Intermission--Sample Student Voice:

Deidra K. Varona / English 090-525 / Nov. 16, 1984

Money Isn't As Important As You Think

If I had a million dollars, the first thing I would buy is a big house fully furnished. Next I would buy a red Mercedez Benz GLE with all the extras. I would buy myself emeralds and diamonds, everything money could buy.

Wait a minute! I just got lost in a million dollars I never had. I lost all contact with reality for a minute. When I imagined all those dollars floating around, I dropped everything. I forgot about the two most important things: my family and my health. Without them I'm nothing, going nowhere.

I should know better than to let money take me away from the people who love me. I have seen this happen to one of the best people in the world, my Uncle Eddie. He has a lot of money now. He has a car, but it is a very old one. He wears expensive clothes, but they are ancient and out of style. I don't think he has bought a new outfit in twenty years. He keeps all his money in the bank and won't touch it. In fact, we don't see him much any more because he thinks we will ask him for money. When he does come, he carries $5.00 or less though he has millions in the bank.

Uncle Eddie is 43 years old and lives alone in a condominium in Manhattan. It isn't as glamorous as it sounds. From what I hear, it has only the furniture that he inherited from his great grandmother. Back when he was earning only $25,000, he was so kind and generous. He had a small, metallic-blue Toyota, and he wore good designer clothing from Sassoon and Bally shoes. He always had a pocket full of money and used to give me an allowance. Christmas time was the best time. We all waited for our Uncle Eddie and when he came to our house he had many gifts for each one of us. Back then he was poor compared to now. The funny thing is that now that he doesn't come any more, we do not miss him.

The money that my uncle inherited transformed him. It made him greedy, crazy. Yet I love him dearly and feel compassion for him. To see greed and selfishness take possession of him has changed my own attitude toward money. I never want to forget my loved ones in my family. I always want to be in the best of health. Family and health are more important to me than money any day. I wish that my Uncle Eddie could figure that out while he is still healthy and has time. It's never too late to re-learn.

Intermission -- Sample Student Voice:

Tien Xuan Ho / English 090-991 / March 22, 1986 / #3 VISUAL

THE ROOMS

Every night I have to read the sign on the door: "Please open and close the door easy and walk softly on the steps." As I turn the key in the lock, it is always quiet, and no matter how soft my steps are, I always make cracking noises while walking up. Sometimes the hallway is dark, for there could be nobody around to turn the light on at dusk. Another turn of the lock and I listen to hear if there are any noises inside. If yes, I feel a little down and if not--relief.

Once inside, I hang up my coat in the closet and proceed into the kitchen which is the first room as I enter. The woman is preparing for her husband and her brother-in-law. I say hello to her. Her brother-in-law, like always, is sitting at the small table in the big living room and watching TV.

Half of the living room is in darkness because the lamp is so small and the space around is so vast. There is no sofa, no love-seat, just a small table and five or four chairs around it. A cabinet in the corner holds a couple of pieces of silverware. In the dim light of the room one could take them for antiques. I once asked the woman if they were. She said no, they were new and for decoration. On top of the cabinet is a cheap stereo component with two detachable speakers. Around them are tapes of contemporary music and Vietnamese music. When Vietnamese music is played, I can't help a feeling of nostalgia. A little bit above the cabinet is a clock which tells surprisingly accurate time. A pot of plastic flowers is placed by the left speaker. In the corner is a mobile stand with a 13" black and white TV set on it and next to it is a computer game set.

I walk through the bedroom of the woman's brother-in-law, then through her room and I enter my room. There's no other way. I have been sharing the apartment with them for almost four months and I don't think it's a problem anymore as I used to when I first moved in. All I need is a little alertness and a little common sense to know when not to walk through their rooms.

As soon as I close the door of my bedroom behind me, I throw the bag in the corner, get undressed and lie down. The room smells of dirty socks. Now I can have my rest after a long exhausting day. I can hear the noises made by the woman in the kitchen, the hissing sound of boiling water, the noises from the TV her brother-in-law is watching.

The room is so small that the bed almost eats up the whole floor, leaving no space for anything else except a little cabinet at the corner. The bed is made up with two mattresses put on top of each other. On it are two pillows and a multicolor blanket. I haven't washed this blanket for almost six months. I do everything in bed. I lie in it, sit in it, read in it, eat in it, write in it, and dream of a palace all for myself.

White walls surround me. On the one on the left of my bed is a piece of paper with two words in bold letters on it that say: SAFE, SANE. It is to remind me to stay cool when I am in my room all alone and high. Just below it are Sade and Picasso's portraits. They are looking at me from their dwellings on the wall. Both are cut-outs from magazines. I look at Sade and I want to bite up her erotic mouth. Right by Sade is a little abstract painting with no signature on it and above it is a notice that says I have an appointment with my dentist on March 25, and beside it is a Christmas card from a friend in Australia. A little bit above the foot of the wall are three prints by a German artist. One shows a little studio with red light coming from a lamp and a guy, maybe the artist himself, is on his knees in front of a giant portrait of Mao Tse Tung. Another one shows a crazy festival of colors with two dancing figures that stand out in yellow, and their heads are two political symbols.

I have a small cabinet at the corner where I put my books on the top shelf, blue jeans on the second shelf and papers of all kind on the third shelf. On the bottom of the cabinet are socks, dirty and clean. Shirts and jackets are hung on the racks right by the cabinet and on the floor are more shirts and four pairs of shoes, a Walkman, a couple of tapes, a pocket radio, half-filled glasses, a candle, cigarettes, books of matches, pens , books and newspapers.

The room has a window with brown curtains. I usually sit by this window to watch people walking and listen to the noises of the city. Sometimes on clear nights I have the room soaked in darkness while I lie in bed, watching a piece of moonlight on the floor and see it shortens as the moon makes its way up in the sky. The door of my room opens to the bedroom of the woman and her husband. I always shut it real tight and can never hear anything coming from there.

I have about 15 days left in this house; I am moving out at the end of this month. The rent of my next place will be higher but at least I won't have to walk through anybody's bedroom to get to the toilet.

Donald Macaluso / English 090-521 / April 23, 1987 / #2

VISION OF A CITY

There are different ways of seeing what is happening around us. I work for New York City Sanitation. The job I perform is one that is clearly needed. What is not clear to most native New Yorkers is how the city looks to the people assigned the task of keeping the Big Apple clean. I am going to show people what the city looks like to a sanitation worker. I would like to say that when someone works outside, vision is not just with the eyes, it is with all the senses.

It is 6:45:AM. I just arrived at work. The garage is full of my co-workers. Everyone has that "wish I could have stayed in bed" look. I feel the same way! Today my assignment is collection. Collection is the normal refuse pick-up that most home owners are accustomed to. By the time I change into uniform and go downstairs, it is 7:00AM. The twelve to eight foreman Pattie Odessa is going to start roll call. As I look around I see a lot of faces, expressing that a long day is about to begin. My partner for today is Willie Night. Willie and I get along well. Willie is about five-ten, two hundred twenty pounds. He is built like a truck. His arms are twice the size of mine. As we check our truck out for a possible mechanical failure I tell Willie today we will be a good "salt and pepper" team. He answers quickly, in a deep voice, "No sweat Mack. We'll get it up and then relax." By the way, Willie is Black and I am white. When we had three people to a crew, two blacks and a white, we made an "oreo."

As we are driving to meet the section foreman I notice that the city is just starting to come awake. There are only a few people at each bus stop. Traffic is still fairly light, but it definitely will not last long. On the way to the section headquarters we stop for coffee. Coffee to go, that is. We arrive to the section about 7:20AM. Foreman Richie Terrana gives us our route for the day. Willie and I take a broom and shovel. We will need it to clean the mess when a plastic bag breaks. A real pain in the ass. Our route is 22nd, 21st, and 20th Streets, from 1st Avenue to Park Avenue. Then the bag stops, around Peter Cooper Village.

The truck will be loaded when we finish all the streets. We will pick them up first to avoid traffic later. At 7:30AM we arrive at 20th Street. My heart is starting to pump faster in anticipation of the work ahead. One can never know how long a city block looks to my eyes at this time, but the only way to finish, is to start. The first stop is about ten cans all over-flowing. The can without a cover securely on I give a good kick, this will scare out any rats or cats that might be inside. This time no furry critters, "Thank God!" I start emptying the cans into the hopper of the truck. "Beep, beep," a car is now in back of us. The driver wants us to stop and go around the block, so he can pass. If we did this, we would never get done. There is just too much traffic in the city. When we finish the block, the driver of the car is very happy to pass us. I wish I could take a picture of his face. If looks could kill, I'd be dead. Willie and I are already starting to sweat. It is about sixty-five degrees and sunny outside. The next two blocks are worse. There are a couple of cars double parked. We have to carry about twenty bags an extra twenty or thirty feet. The same applies to the next two stops. I jog a bit, and after just the first line of the route I feel like I went over a mile. I hope I can keep up with Willie who is not showing any signs of slowing down.

In the middle of Twenty-first Street a line of cars are waiting to pass. The horns are starting to blow. I can feel the blood pressure rising in the anxious drivers who think they may be late for work. The air is cool but the pile of bags at Baruch College is still ten feet long and four feet high. I glance up at the line of cars, about ten cars long. I am out of breath and the bags are feeling heavier. The horns blow again. I answer, under my breath, with a few, two to be exact, choice words. Finally everyone is happy. The stop is completed. Willie is the driver today. He moves the tuck so traffic can pass. The driver of the first car is not sure he can make it by. We wait, and he still dose not move. I wave him on and still he doesn't want to chance scratching his car. That's all I can do. I have work to finish. Willie backs into the next stop. Then all hell breaks loose. Four other drivers get out of their cars to yell at the driver of the first car for not moving. Willie and I watch as we work. We do not want to witness or get involved in anything. Well luckily no one threw any punches.

[cont'd on next page]

Now we are on Twenty-second Street. The truck is starting to smell a little. It will get worse as the weather gets hotter. Our first stop is ashes. I hate ashes. The dust gets in my eyes, nose, mouth, and hair. It leaves a coating all over me. Sometimes the porter who brings out the ashes will wet the cans down. This will keep the dust from flying around, but the cans are much heavier. A line of cars is starting to form again. Believe it or not, the first car is the same one as before, the one who would not move. I have no idea why he is back, and I do not wish to know. The rest of the block is just as tough. As we leave the ash stop, there is a car waiting in front of the truck. He wants Willie to back up so he can park. I politely explain that the truck can not back up due to the cars in back of it. The driver doesn't look too happy. I think he's taking this personally. Well there's nothing I can do about it. In the middle of the block there is a pretty big bag stop. The bags are all blocked by parked cars. We have to drag each bag out. The bags are getting heavy now. As we load them into the truck, cars are whizzing by. With the bags finished there are just a few can stops ahead, until First Avenue. The truck is just about loaded. Finally the last can is dumped. Willie and I close the hopper so nothing will fall out, and off to the dump we go.

The dump is on Fourteenth Street and the West Side Highway. We will go west along Twenty-third Street. It is just about nine a.m. The streets are crowded now. People are buzzing around like ants. I assume everyone is going to work. Traffic is heavy and we are stuck fighting it, cross-town no less. Along the way I girl-watch. It is fun getting different reactions from the women, while on the way to the dump. Most do not return my energetic "Good Morning". The rest of the ride goes with out incident. At the dump the truck is weighed to see how much we picked up. We have picked up seven tons, not bad. The lighting at the dump is okay. Even so, it is ominous looking, always kind of grayish. The air is always filled with dust from the trucks dumping their loads. I can feel and see the dust in the air. I try not to breathe as hard as usual. As we leave the dump, I am on top of a ramp, looking over New York Bay. The sun is trying to shine, and the water sparkles a little, reflecting the sun. Seagulls are feeding on the refuse in a loaded barge to the side of the dump. The World Trade Center looks towering from here. I wonder how they built it.

Back across Twenty-third Street and girl-watching again, I notice a young lady dressed in a red skirt and jacket with a white blouse. I give my usual "Good Morning!" She returns it with a big smile and good morning too. I smile and tip my hat. At that moment the truck pulls out from the light. It is a good feeling when New Yorkers are friendly, especially the ladies. The rest of the ride is normal, cut off by about half a dozen cabs and people, trying to cross the street.

About nine-thirty we reach our first bag stop. Willie and I have to load the truck again. By this time we are cooled down. I had to put my sweat shirt back on. The sweat on my t-shirt was making me cold. Now it's time to start over. There are only about seven stops but all are very big. These bags contain compacted material. Each bag may weigh fifty to two hundred pounds. People put anything and everything in these bags. I must be very careful picking up each bag, not to use my body. Any glass or other sharp object will cut deep with the weight of the bag behind it. When we finish the first stop I am warm and starting to sweat again. As I pick up each bag I can feel my arms, legs and wrists cracking. Boy, is this a workout! The worst is being all set to lift a bag, and as it just clears the ground, the bottom breaks. I spring up from the reduced weight, and have a mess to sweep up.

All in all the route is completed. Willie and I are tired but all that is left is to dump the truck. By quitting time we ride to the dump and return to the garage. The shower feels great. It gives me new energy. At three o'clock everyone signs out, to go home. Tomorrow is another day.

Everyone knew this was a good day. The weather was good. There was no rain, hail, or snow hitting us in the face. There was no cold wind to numb our toes, feet, hands, and teeth. The weather can be our best friend or worst enemy. Next time that it's rainy, cold, and windy, take a look outside, and try to imagine working under those conditions.

Intermission--Sample Student Voice:

Fernando Garcia / English 090-422 / April 1985

The Best Friend

The best friend is said to be a dollar in your pocket because when you haven't a cent to your name, you haven't a friend either. It think this is true in a way, but there's always someone who is liable to stick to your side, like Jimmy.

Jimmy and I had been best friends for many years. We were little kids when we met, and I don't remember where or when it all started for us, but I knew it was special because we could read each other like books even before we started our friendship. My twin brother Orlando was Jimmy's friend, too. I guess we knew a true friend when we saw one. The three of us were always together, getting in trouble, playing ball, anything to stay busy--and that we did.

About four years ago I had to move with my family to Puerto Rico. We stayed in a small town called Toa Alta. I really didn't like it there, so I kept wishing that I was back in New York with Jimmy. I didn't like Toa Alta because it was too quiet for me. I like action.

During this time I didn't hear from Jimmy and I didn't bother to write either. I met many people my age, but I didn't trust them. I don't think I was trying to compare them to Jimmy, but I just didn't feel free with them as when I was with him. Four years went by before I graduated from Papa Juan High and returned from Toa Alta to New York where my best friend was still living.

When we met again it was different. I could feel the tight nervousness riding on the words as we spoke. We kept our guards up, you could say. I didn't like it because we had never been like that; we had felt secure together in the past. So we spoke of the past years. Everything we said revolved around the "good old days." Then our conversation was over and we went our ways. I went home, and he went to catch his train. While walking, I couldn't stop thinking about how much we had changed in thought and attitude. It really dawned on me that we weren't best friends any more. That was all in the past.

Since that encounter I've seen Jimmy only about twice and we hardly spoke. I wonder what had changed us. Was it growing up that had finished our friendship? We can't communicate now, and it breaks my heart that my old friend was lost somewhere in time.

Goran "George" Moberg is a teacher and consultant on writing. Before settling in New York, he studied at the Universities of Stockholm and Paris. He has a Ph.D. in English from Columbia University.

An award-winning writer, he has published poems, articles, and textbooks. He is now at work on ***The New English Teacher in America***, a book for the general public about new developments in the pedagogy of English studies. Though he uses a yellow pad for difficult spots, he does most of his writing on a 286 computer, with the aid of a laser printer; in fact, this book was typeset with that equipment.

Dr. Moberg has taught writing at Colorado State University, Wilmington College (Ohio), and the University of Puerto Rico. He is currently an Associate Professor of English at Borough of Manhattan Community College of the City University of New York where he helps his students to compose their essays on an IBM local area network, using Bank Street Writer, Word Perfect, Realtime Writer, and other programs.